112

INTELLIGENT MACHINES

INTELLIGENT MACHINES:

An Introductory Perspective...

of Artificial Intelligence

and Robotics

WILLIAM B. GEVARTER

NASA Ames Research Center, Moffett Field, CA

PRENTICE-HALL, INC., Englewood Cliffs, New Jersey 07632

Library of Congress Cataloging in Publication Data

Gevarter, William B. (date)
  Intelligent machines.

  Bibliography: p.
  Includes index.
  1. Artificial intelligence. 2. Robotics. I. Title.
  Q335.G483 1985   001.53'5   84–22326
  ISBN 0-13-468810-4

Editorial/production supervision and
  interior design: David Ershun
Cover design: Lundgren Graphics, Ltd.
Manufacturing buyer: Gordon Osbourne

This book has its roots in a series of NASA publications by
Dr. Gevarter while at NASA Headquarters. However, this
book is not sponsored by NASA and the opinions and
conclusions in it are not necessarily those of NASA.

Printed in the United States of America

10 9 8 7 6 5 4 3 2

ISBN 0-13-468810-4   01

Prentice-Hall International, Inc., *London*
Prentice-Hall of Australia Pty. Limited, *Sydney*
Editora Prentice-Hall do Brasil, Ltda., *Rio de Janeiro*
Prentice-Hall Canada Inc., *Toronto*
Prentice-Hall Hispanoamericana, S.A., *Mexico*
Prentice-Hall of India Private Limited, *New Delhi*
Prentice-Hall of Japan, Inc., *Tokyo*
Prentice-Hall of Southeast Asia Pte. Ltd., *Singapore*
Whitehall Books Limited, *Wellington, New Zealand*

To my wife Annette

CONTENTS

PREFACE

This book is intended to fill the void that now exists between very technical treatises on artificial intelligence and robotics, and the "gee-whiz" approach of the popular magazine and newspaper cover stories and articles. Its focus is on creating an overview (the forest), but it incorporates enough conceptual material to provide an introductory understanding of how the individual systems (the trees) are produced. The intent is to cover the following:

- Basic concepts
- Component systems
- The state of the art
- Current research directions, organizations, and funding sources
- Present and future applications
- Future directions
- Sources for further information

The goal is to provide to the intelligent layman, the engineering manager, and technical people from other fields with a grasp of the basic concepts, structure and activities of these newly emerging technologies. The objective is to provide an integrated perspective from which to evaluate intelligently what is going on in order to make appropriate decisions about future activities and to help provide direction for individual programs.

As a result, the book can also be used as a textbook for an introductory course in artificial intelligence and/or robotics, serving as a readily understandable tutorial and providing an overall context and technology assessment.

To facilitate the use of this book by those readers who need information on a specific topic area, each chapter has been made reasonably self-contained.

William B. Gevarter

ACKNOWLEDGMENTS

I wish to thank the many people and organizations of the artificial intelligence and robotics communities who have contributed to this book by providing information and source material and by reviewing portions of the book and suggesting corrections, modifications, and additions. However, the responsibility for any remaining errors or inaccuracies (for which I apologize in advance) must remain with the author.

I would particularly like to express my thanks to NASA and the National Bureau of Standards for providing me with the opportunity to write the seven-volume NBS/NASA *Overview of Artificial Intelligence and Robotics* series. This book can be considered a synthesis, extension, and update of that series.

This book would not have been possible, of course, without the efforts of those inspired individuals who have created the theory, concepts, and systems upon which these fields are based. I am deeply grateful to them for the service they have rendered.

I would particularly like to thank James Albus of the National Bureau of Standards, Robert Hong and his associates at Grumman Aerospace Corporation, Jerry Cronin of the U.S. Army Signal Warfare Lab., Jude Franklin of the U.S. Navy Center for Applied Research in Artificial Intelligence, and Commander Ronald Ohlander of DARPA for their encouragement and aid in this project.

However, I would indeed be remiss if I did not mention Mrs. Margie Johnson, who performed heroically in typing and facilitating the publication of the original NBS series of reports.

Finally, I would like to express my deep appreciation to my wife for not only typing much of the manuscript but for never failing to encourage me in the development of this book, despite the serious time inroads it made in our lives.

It is not the intent of the author or Prentice-Hall to recommend or endorse any of the manufacturers or organizations named in this report, but simply to attempt to provide an overview of the artificial intelligence and robotics fields. However, in such diverse and rapidly changing fields as AI and robotics, important activities, organizations, and products may not have been mentioned. Lack of such mention does not in any way imply that they are not worthwhile. The author would appreciate having any such omissions, oversights, or needed corrections called to his attention so that they can be considered for future editions.

Part I

ARTIFICIAL INTELLIGENCE

1

ARTIFICIAL INTELLIGENCE: WHAT IT IS

1-1. DEFINITIONS AND APPLICATIONS

1-1.1. Definition

*Artificial intelligence** (AI) is an emerging technology that has recently attracted considerable publicity. Many applications are now under development. One simple view of AI is that it is concerned with devising computer programs to make computers smarter. Thus research in AI is focused on developing computational approaches to intelligent behavior. This research has two goals: (1) making machines and computational processes more useful and (2) understanding intelligence. This book is concerned primarily with the first goal.

The computer programs with which AI is concerned are primarily symbolic processes involving complexity, uncertainty, and ambiguity. These processes are usually those for which algorithmic solutions do not exist and search is required. Thus AI deals with the types of problem solving and decision making that human beings continually face in dealing with the world.

This form of problem solving differs markedly from scientific and engineering calculations that are primarily numeric in nature and for which solutions are known that produce satisfactory answers. In contrast, AI programs deal with words and concepts and often do not guarantee a correct solution—some wrong answers being tolerable, as in human problem solving.

Table 1-1 provides a comparison of AI and conventional computer programs. A key characteristic of AI programs is *heuristic search*. Boraiko (1982, p. 448)

---

*Also sometimes referred to as *machine intelligence* or *heuristic programming*. Section 1-4 expands on the definition of AI given in this section.

quotes Minsky as saying: "If you can't tell a computer how best to do something, program it to try many approaches." However, in complex problems the number of possible solution paths can be enormous. Thus AI problem solving is usually guided by empirical rules—rules of thumb—referred to as *heuristics*, which help constrain the search.

TABLE 1-1   Comparison of AI with Conventional Programming

Artificial Intelligence Programming	Conventional Computer Programming
Primarily symbolic processes	Often primarily numeric
Heuristic search (solution steps implicit)	Algorithmic (solution steps explicit)
Control structure usually separate from domain knowledge	Information and control integrated
Usually easy to modify, update, and enlarge	Difficult to modify
Some incorrect answers often tolerable	Correct answers required
Satisfactory answers usually acceptable	Best possible solution usually sought

Another aspect of AI programs is the extensive use of *domain knowledge.* * Intelligence is heavily dependent on knowledge. This knowledge must be available for use when needed during the search. It is common in AI programs to separate this knowledge from the mechanism that controls the search. In this way, changes in knowledge require only changes in the knowledge base. In contrast, domain knowledge and control in conventional computer programs are integrated. As a result, conventional computer programs are usually difficult to modify, as the implications of the changes made in one part of the program must be carefully examined for the impacts and the changes required in other parts of the program.

1-1.2. The Basic Elements of AI

Nilsson (1981/1982; see also Brown and Cheeseman, 1983), a pioneer in AI and at the time of this writing head of the Stanford Research Institute (SRI) AI Center, likes to characterize the components of AI in terms of what he calls the *onion model* (See Figure 1-1). The inner ring depicts the basic elements from which the applications shown in the next ring are composed. We will consider first the quadrant designated as heuristic search.

Heuristic search.   Much of the early work in AI was focused on devising programs that would search for solutions to problems. Note that every time one makes a decision, the situation is changed, opening up new opportunities for further decisions. Therefore, there are always branch points. Thus one of the usual ways of representing problem solving in AI is in terms of a tree (see, e.g., Figure 3-3, Chapter 3), starting at the top with an initial condition and branching every time a decision is made. As one continues down the tree many different decision possibilities open up, so that the number of branches at the bottom can get to be enormous for

*Knowledge of the problem area of interest.

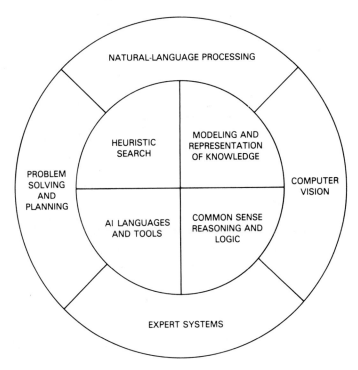

Figure 1-1   Elements of AI.

problems requiring many solution steps. Therefore, an efficient way is needed to search such trees.

Initially, there were "blind" methods for searching trees. These were orderly search approaches that assured that the same solution path would not be tried more than once. However, for problems more complex than games and puzzles, these approaches were inadequate. Therefore, rules of thumb (empirical rules), or heuristics, were needed to aid in choosing the most likely branches, so as to narrow the search. As an example, a simple heuristic to help choose which roads to follow when driving in the evening on back roads from Washington, DC, to San Francisco is: "Head for the setting sun." This may not produce the optimum path but can serve to help advance one toward one's goal. Heuristic rules like this can help guide search—reducing search enormously.

Knowledge representation.   AI researchers discovered that intelligent behavior is not so much due to the methods of reasoning as it is dependent on the knowledge one has to reason with. (As human beings go through life they build up tremendous reservoirs of knowledge.) Thus when substantial knowledge has to be brought to bear on a problem, methods are needed to model this knowledge efficiently so that it is readily accessible. The result of this emphasis on knowledge is that knowledge representation is one of the most active areas of research in AI today. The needed knowledge is not easy to represent, nor is the best representation obvious for a given task.

Commensense reasoning* and logic.   AI researchers found that common sense (virtually taken for granted in human beings) is one of the most difficult things to model in a computer. It was finally concluded that common sense is low-level reasoning, based on a wealth of experience. In acquiring common sense we learn to expect that when we drop something it falls,[†] and in general what things to antici-pate in everyday events. How to represent common sense in a computer is a key AI issue that is unlikely to be completely solved in the near future.

Another area that is very important in AI is logic. How do we deduce some-thing from a set of facts? How can we prove that a conclusion follows from a given set of premises? Computational logic was one of the early golden hopes in AI to provide a universal problem-solving method. However, solution convergence proved to be difficult with complex problems, resulting in a diminishing of interest in logic. Logic is now enjoying a revival based on new formulations and the use of heuristics to guide solutions.

AI languages and tools.   In computer science, specific high-level languages have been developed for different application domains. This has also been true for AI. Currently, LISP and PROLOG are the principal AI programming languages. To date, LISP (List Processing Language, developed in the late 1950s by John McCarthy, then at MIT) has been the prime language for AI in the United States. Utilizing LISP, software tools have been devised for expressing knowledge, formulating ex-pert systems, and basic programming aids. (More detailed information on LISP and PROLOG, together with examples, are given in Appendix C.)

1-1.3. Principal AI Application Areas

Based on these basic elements, Nilsson identified four principal AI applica-tion areas (shown in the outer ring of Figure 1-1).

Natural language processing (NLP).   NLP is concerned with natural language front ends to computer programs, computer-based speech understanding, text understanding and generation, and related applications. A detailed overview of NLP is given in Chapter 9.

Computer vision.   Computer vision is concerned with enabling a computer to see—to identify or understand what it sees, to locate what it is looking for, and so on. A detailed overview of computer vision is given in Chapter 8.

Expert systems.   This is perhaps the "hottest" topic in AI today. How do we make a computer act as if it were an expert in some domain? For example, how

---

*Commonsense reasoning is informal "everyday" reasoning.

[†] Such knowledge is sometimes referred to as *commonsense physics.*

do we get a computer to perform medical diagnosis or VLSI design? A detailed overview of expert systems is given in Chapter 6.

Problem solving and planning.   There is a need for general-purpose problem-solving approaches to attack problems for which there are no experts. In addition, there are some basic planning systems that are more concerned with solution techniques than with knowledge. A comprehensive overview of problem-solving and planning techniques is given in Chapter 3, and planning systems are discussed in Chapter 7.

1-1.4. Is AI Difficult?

The popular view that the study of AI is difficult is due partially to the awe associated with the notion of intelligence.* It is also due to the nomenclature used in AI and to the large size of some AI computer programs. However, the basic ideas of AI are readily understandable, even though in complex applications, the "bookkeeping" associated with such programs can be arduous. Before we go into details on these basic ideas, it is illuminating to consider mechanization and automation and the relationship of AI to them.

1-2. MECHANIZATION AND AUTOMATION

To understand better what is meant by *artificial intelligence* and *robotics*, it is helpful to step back a bit and look first at terms such as *mechanization* and *automation*. To do this we will try to synthesize the views of others who have approached this problem.

The original industrial revolution was based on mechanization. Mechanization was the use of machines to take over some of the previous muscle jobs performed by either animals or human beings. Laurie (1979) states:

> When we apply ordinary production techniques—the application of leverage and power—to a process, we are mechanizing it. Automation involves a good deal more. . . . Automated devices are truly automated when feedback information automatically causes the machinery to adjust to reachieve the norm. The internal adjustments of the machine or system are made by servomechanisms [P. 355]

> Automation is the achievement of self-directing productive activity as a result of the combination of mechanization and computation. . . . [P. 15]

Peter Marsh (1981, pp. 419–420) elaborates further on mechanized machines, automatic devices, and automated devices:

*Indeed, researchers cannot even agree on a definition of intelligence itself. One paradoxical definition is: "Intelligence is what an intelligence test measures."

[The classification of mechanization] depends on whether machines or combinations of animals and people are responsible for the three fundamental
elements that occur in every activity (human or otherwise)—power, action
and control. [Simple mechanized devices] need a human to control them. If a
mechanical device is responsible for control, however, we have a self-acting or
automatic device. Automatic devices are not the same as automated ones . . .*
automation equals mechanization plus automatic control plus one (or more)
of three extra control features—a "systems" approach, programmability, or
feedback.

Extras That Make Automation

With a systems approach, factories make parts by passing them through suc-
cessive stages of a manufacturing process without people intervening. Thus
the transfer lines of car factories in the 1930s count as automated systems.

    With programmability—the second of the three "extras" that define
automation—an automated system can do more than one kind of job. Hence
an industrial robot is an automated, not an automatic, device. The computer
that controls it can be fed different software to make the machine do different things—for example, spray paint or weld bits of metal together. Finally,
[external] feedback makes an automatic machine alter its routine according
to changes that take place around it. An automatic lathe with feedback—in
which, for instance, a sensor detects that the metal it is cutting is wrongly
shaped and so instructs the machine to stop—is thus an automated device. It
is clearly more useful than a lathe without this feature.

1-3. TOOLS, MACHINES, TELEOPERATORS, AND ROBOTS

To extend the concepts of mechanization and automation further, we will consider
tools, machines, teleoperators, and robots. To do this, we will utilize Marsh's (1981)
basic elements: power, action, and control.

- *Tool:* A device used to perform an action. If used by a human being, the person provides the power and control.
- *Machine:* A device that utilizes nonhuman power to do an action. For a simple
  machine the human being provides the control.
- *Teleoperator:* A machine capable of action at a distance under the control of
  a human being.
- *Robot:* A flexible machine capable of controlling its own actions for a variety
  of tasks utilizing stored programs. Basic task flexibility is achieved by its
  capability of being reprogrammed. More advanced—intelligent—robots would
  be capable of setting their own goals, planning their own actions, and correcting for variations in their environment.

*The term *automatic* has a somewhat different connotation in artificial intelligence, as,
for example, in the term *automatic programming.*

1-4. FURTHER DEFINING AI

Before examining the relationship of artificial intelligence to automation, it is helpful to enlarge on our definitions of AI.

Laurie (1979, p. 15) defines a computer as "an electronic device capable of following an intellectual map. We call the map a program." Arden (1980, p. 9) suggests that "computer science is the study of the design, analysis, and execution of algorithms* in order to better understand and extend the applicability of computer systems."

Although everyone agrees that artificial intelligence is difficult to define precisely, the most commonly accepted definition is that "artificial intelligence is the branch of computer science devoted to programming computers to carry out tasks that if carried out by human beings would require intelligence."

A slightly different definition is given by Duda et al. (1979, p. 728):

Artificial intelligence (AI) is the subfield of computer science concerned with the use of computers in tasks that are normally considered to require knowledge, perception, reasoning, learning, understanding and similar cognitive abilities. Thus, the goal of AI is a qualitative expansion of computer capabilities.

Nilsson (1980, p. 2) states:

AI has also embraced the larger scientific goal of constructing an information-processing theory of intelligence. If such a science of intelligence could be developed, it would guide the design of intelligent machines as well as explicate intelligent behavior as it occurs in humans and other animals. Since the development of such a general theory is still very much a goal, rather than an accomplishment of AI, we limit our attention here to those principles that are relevant to the engineering goal of building intelligent machines.

More recently, Nilsson (1981/1982) indicated that he would like to narrow the working definition of AI even further to the central processes of intelligence. He thus states:

With regard to humans, I am inclined to consider as *central* those cognitive processes that are involved in reasoning and planning. Work on automatic methods of deduction, commonsense reasoning, plan synthesis, and natural-language understanding and generation are examples of AI research on central processes.

Perhaps as important as the processes themselves is the "knowledge" they manipulate. In fact, the subject of knowledge representation formalisms is a good starting point for a more detailed explanation of just what I think AI is.

*An *algorithm* is a set of rules or processes for solving a problem in a finite number of steps.

Arden (1980, pp. 22 and 23) observes:

Though "intelligent behavior" is difficult to define, and is currently understood differently by different people, there has been some convergence of views within the AI community as the technical requirements for the computer solution of certain classes of problems becomes better understood. To be sure, the human solution of a complex equation might be classified as intelligent behavior, while the corresponding action by a machine might not be so classified, even though both machine and man had been programmed for (learn) the process. One possible requirement is that there be something unstructured, something nondeterministic, for the solution process to qualify as intelligent. Another is that it depends on the knowledge that must be used in obtaining the solution, or on the methods used. . . .

Another important aspect is the use of heuristic rules* of the kind humans use to solve problems. Although, in general, such rules cannot be proved effective, they often lead to solutions. Some computer scientists argue that heuristic programming better describes the field now called "artificial intelligence."[†]

Hayes-Roth (1981, p. 1) notes that:

AI provides techniques for flexible, non-numerical problem-solving. These techniques include symbolic information processing, heuristic programming, knowledge representation, and automated reasoning. No other fields or alternative technologies exist with comparable capabilities. And nearly all complicated problems require most of these techniques. Many forces combine to identify AI as the central technology for exploitation. Systems that reason and choose appropriate courses of action can be faster, cheaper, and more effective and viable than rigid ones. To make such choices in realistically complex situations, the system needs at least rudimentary understanding of mundane phenomena.

In summary, AI is concerned with intelligent behavior, primarily with nonnumeric processes that involve complexity, uncertainty, and ambiguity and for which known algorithmic solutions do not usually exist. Unlike conventional computer programming, it is knowledge based, almost invariably involves search, and uses heuristics to guide the solution process.[‡]

Thus AI can be considered to be built on the following:

Heuristics are rules of thumb (compiled experience) used to help guide problem solving. They do not necessarily guarantee a solution, as algorithms do.

[†]Reprinted by permission from B. W. Arden, ed., *What Can Be Automated*. Copyright 1980 by MIT Press, Cambridge, Mass.

[‡]As AI matures, the expectations associated with it are increasing. Schank (1983) states that it is time to demand learning capability from AI programs. He thus suggests a new definition: "AI is the science of endowing programs with the ability to change themselves for the better as a result of their own experiences."

1. Knowledge of the domain of interest.
2. Methods for operating on the knowledge.
3. Control structures for choosing the appropriate methods and modifying the data base (system status) as required. This contrasts with conventional computer programs, which utilize known algorithms for solution, are primarily numeric (number crunching) in nature rather than symbolic manipulation, and in general do not require knowledge to guide the solution.

1-5. RELATIONSHIP OF AI TO AUTOMATION

Artificial intelligence can be considered to be the top layer of control on the hierarchical road to autonomous machines. This is illustrated in Fig. 1-2.

However, AI includes a large area of activity which is not normally included in automation, for example:

- Natural language processing
- Perception
- Intelligent information storage and retrieval
- Game playing
- Automatic programming
- Computational logic
- Problem solving
- Expert systems

Nevertheless, as computer-integrated manufacturing and intelligent robots emerge, AI will have a major role to play. AI contributions to perception and object-oriented programming are reviewed by Brady (1984) for this new breed of robots.

1-6. AI AND OTHER FIELDS

Duda et al. (1979, pp. 729-730) state:

> Historically, AI has both borrowed from and contributed to other closely related disciplines concerned with advanced methods for information processing. Thus, links exist between AI and aspects of such theoretical areas as mathematical logic, operations research, decision theory, information theory, pattern recognition and mathematical linguistics. In addition, research in AI has stimulated important developments in software technology, particularly in the area of advanced programming languages. What distinguishes AI from these related fields, however, is its central concern with all of the mechanisms of intelligence.

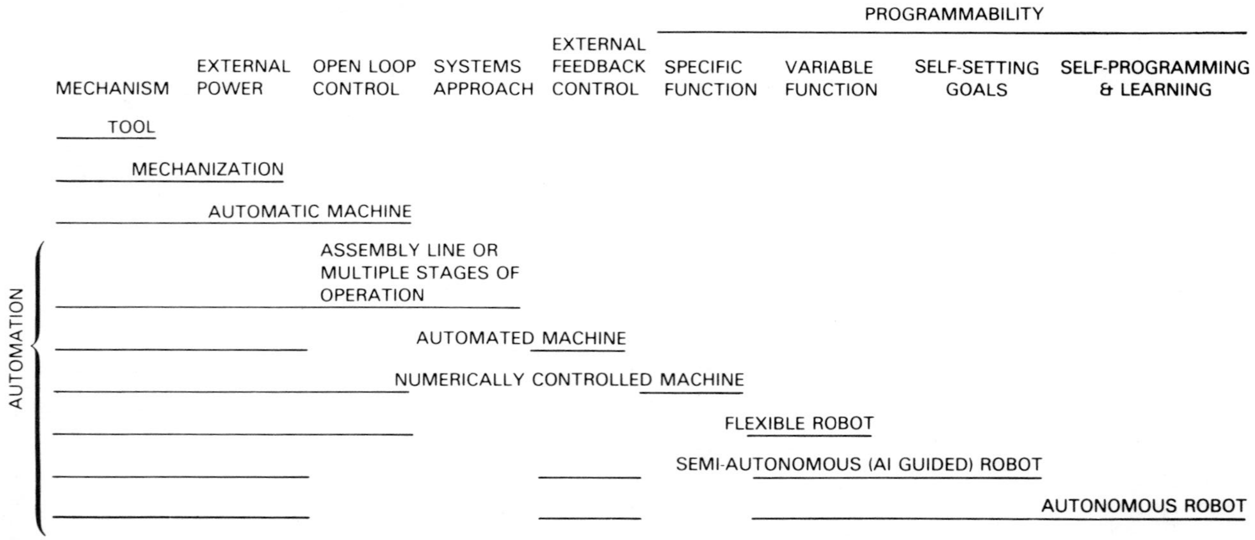

Figure 1-2 Mechanization, automation, and artificial intelligence. (Derived from Marsh, 1981.)

REFERENCES

Arden, B. W. (Ed.), *What Can Be Automated*. Cambridge, MA: MIT Press, 1980.

Boraiko, A. A., "The Chip," *National Geographic*, Oct. 1982, pp. 421–456.

Brady, M., "Artificial Intelligence and Robotics, in *Artificial Intelligence*, M. Brady and L. Gerhardt (Eds.). New York: Springer-Verlag, 1984.

Brown, D. R., and Cheeseman, P. C., *Recommendations for NASA Research and Development in Artificial Intelligence*, SRI Project 4716, SRI International, Menlo Park, CA: April 1983.

Duda, R. O., et al., "State of Technology in Artificial Intelligence," in *Research Directions in Software Technology*, P. Wenger (Ed.). Cambridge, MA: MIT Press, 1979, pp. 729–749.

Hayes-Roth, F., "AI: The New Wave—A Technical Tutorial for R & D Management," Rand Corp., Santa Monica, CA: 1981 (AIAA-81-0827).

Laurie, E. J., *Computers, Automation, and Society*. Homewood, IL: Richard D. Irwin, 1979.

Marsh, P., "The Mechanization of Mankind," *New Scientist*, Feb. 12, 1981, pp. 418–421.

Nilsson, N. J., *Principles of Artificial Intelligence*. Palo Alto, CA: Tioga, 1980.

Nilsson, N. J., "Artificial Intelligence: Engineering, Science or Slogan," *AI Magazine*, Vol. 3, No. 1, Winter 1981/1982, pp. 2–9.

Schank, R. C., "The Current State of AI: One Man's Opinion," *AI Magazine*, Vol. 4, No. 1, Winter–Spring 1983, pp. 3–8.

2

WHAT THIS BOOK WILL COVER

Now that the concept of artificial intelligence has been introduced, we will explore it in greater detail. We begin by examining the foundations of AI: automated problem solving, knowledge representation, and computational logic.* Then, having set the base, we next explore applications: expert systems, planning, computer vision, natural language processing, and speech understanding.† Based on this, we then examine what it all means, including future applications and implications. Next we turn to robotics (in Part II) and engage in similar explorations. Finally, we observe that the two lines of development merge, compounding their impact on the future.

As an aid to the reader, a glossary of AI terms and a list of further sources of information on AI is given at the end of Part II. Appendix D gives a brief history of AI. Appendix E contains a summary of the principal players in the unfolding AI drama.

---

*Appendix C provides information on AI languages, tools, and computers.
†Appendix F covers the related area of speech synthesis.

3

SEARCH-ORIENTED AUTOMATED

PROBLEM-SOLVING AND

PLANNING TECHNIQUES

This chapter provides an overview of search-oriented automated problem-solving and planning techniques. It endeavors to present the basic approaches to automated problem solving at a level where the concepts involved can be readily understood. It also provides an indication of the state of the art and current and future research.

3-1. AI AS A PROBLEM SOLVER

One way of viewing intelligent behavior is as a problem solver. Many AI tasks can naturally be viewed this way, and most AI programs draw much of their strength from their problem-solving components. AI applications that have strong problem-solving components are scene analysis, natural language understanding, theorem proving, task planning, expert systems, game playing, and information retrieval and extraction.

Two important types of problem-solving tasks are (1) synthesizing a set of actions (a plan) to achieve a goal, and (2) deduction. The latter involves deducing (or inferring) conclusions from data or a given set of propositions (applications include theorem proving and information extraction). In this chapter we restrict ourselves to action synthesis, leaving a review of deduction for Chapter 5.

Many tasks can be formulated in terms of the question: Given a goal, how do we achieve it? If direct methods are not available for solution, as is the usual case in AI problems, a search procedure to select from the various possible alternatives is required. Thus, finding efficient search methods is one of the central issues in automated problem solving.

3-2. ELEMENTS OF A PROBLEM SOLVER

All action-synthesis problems have certain common aspects: an initial situation, a goal (desired situation), and certain operators (procedures or generalized actions) that can be used for changing situations. In solving the problem, a control strategy is used to apply the operators to the situations to try to achieve the goal. This is illustrated in Fig. 3-1, where we observe a control strategy operating on the procedures to generate a sequence of actions (called a *plan*) to transform the initial conditions in the situation into the goal conditions. Normally, there are also constraints and preconditions (conditions necessary for a specific procedure to be applied) which must be satisfied in generating a solution. In the process of trying to generate a plan, it is necessary for the problem solver to keep track of the actions tried and the effects of these actions on the system state. Figure 3-2 is a restatement of Fig. 3-1 in which we can view the operators as manipulating the data base (representing the problem status) to change the current situation (system state).

3-3. SEARCH

AI problem solving can often be viewed as a search among alternative choices. It is thus possible to represent the resulting search space as a hierarchical structure called a *tree*, an example of which is shown in Fig. 3-3. (Figure 3-3 is a search tree for the elementary problem of finding the simplest route, from city A to the destination city D, from among the network of roads illustrated by the state graph of Fig. 3-4.)

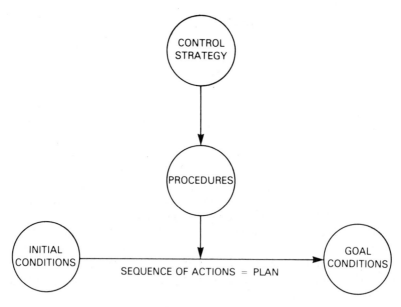

Figure 3-1 Problem solving.

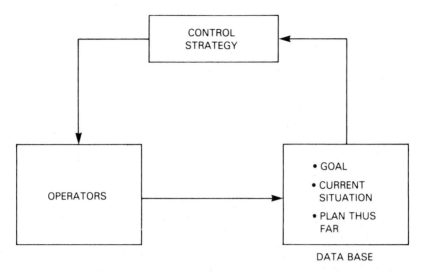

Figure 3-2   Automated problem-solving relationships.

The solution paths run from the initial state (root node) along the branches of the tree and terminate on the leaves (terminal nodes) labeled "goal state."

For a large, complex problem, it is obviously too cumbersome to draw such trees of all the possibilities and examine them directly for the best solution. Thus the tree is usually implicit, the computer generating branches and nodes as it searches for a solution. In searching for a solution we can reason forward as in Fig. 3-3 or backward from the goal (searching an equivalent tree where the root node is the goal).

3-3.1. Blind Search

For fairly simple problems, a straightforward, but time-consuming approach is blind search, where we select some ordering scheme for the search and apply it until the answer is found. There are two common blind search procedures, breadth-first search and depth-first search. In breadth-first search, the nodes of the search tree are generated and examined level by level starting with the root node. In a depth-first search, a new node (at the next level) is generated from the one currently being examined, the search continuing in this way deeper and deeper until forced to backtrack.

Blind search does not make any use of knowledge about the problem to guide the search. In complex problems, such searches often fail, being overwhelmed by the combinatorial explosion of possible paths. If on the average there are n possible operators that can be applied to a node, and the search space is searched to a depth of d, the size of the search space tends to grow in relation to n^d. Heuristic methods have been designed to limit the search space by using information about the nature and structure of the problem domain. Heuristics are rules of thumb, techniques, or empirical knowledge that can be used to help guide search. Heuristic search is one

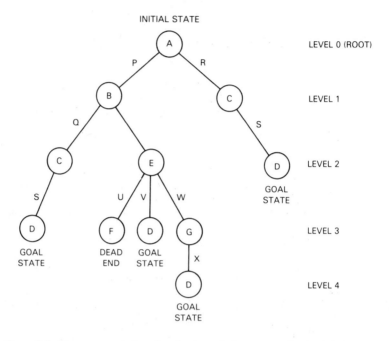

Figure 3-3   Tree representation of paths through the state graph of Fig. 3-4.

of the key contributions of AI to efficient problem solving. It often operates by generating and testing intermediate states along a potential solution path.

One straightforward method for choosing paths by this approach is to apply an evaluation function to each node generated and then pursue those paths that have the least total expected cost. Typically, the evaluation function calculates the cost from the root to the particular node that we are examining and, using heuristics, estimates the cost from that node to the goal. Adding the two produces the total estimated cost along the path, and therefore serves as a guide as to whether to proceed along that path or to continue along another, more promising, path among those thus far examined. However, this may not be an efficient approach to minimize the search effort in complex problems.

Search techniques are now relatively mature and are codified in *The Handbook of Artificial Intelligence* (Barr and Feigenbaum, 1981) and various AI texts. Pearl (1984) provides a good indication of current research in search.

3-4. GAME TREE SEARCH

3-4.1. Representation

Most games played by AI computer programs involve two players making alternate moves. A game representation must thus take into account the opponent's possible moves as well as the player's own moves. The usual representation is a

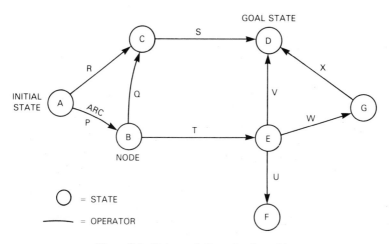

Figure 3-4   State graph for a simple problem.

game tree, which shares many features with a problem reduction representation (discussed in Section 3-5.2). A complete game tree is a representation of all possible plays of such a game.

The root node is the initial state, in which it is the first player's (A's) turn to move. The successors of the root node are the states that A can reach in one move. The successors of these nodes are the states resulting from the other player's (B's) possible replies; and so on. At each play, the players must take into account all the opponent's possible responsive moves. This can be represented by an AND/OR tree. Figure 3-5 is an example of such a tree from the standpoint of player A, who is to move next. Drawn from player A's point of view, A's possible moves under his control are represented by lines leading to OR nodes. These successor nodes are called *OR nodes*, as they represent A's alternative choices. Because A, in arriving at a decision, must also consider all of B's possible moves, B's successor nodes are called *AND nodes*.

3-4.2. The Minimax Search Procedure

The minimax procedure is a strategy for playing a two-person game. According to the minimax technique, player A should move to the position of maximum value to him or her, B responding by choosing a move of minimum value to player A. Given the values of the terminal positions (see Fig. 3-5), the value (shown in parentheses) of a nonterminal position to player A is computed by backing up from the terminals as follows:

- The value of an AND node is the maximum value of any of its successors.
- The value of an OR node is the minimum value of any of its successors.

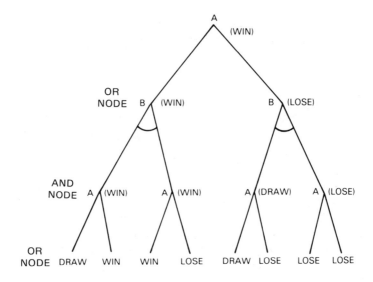

Figure 3-5   A game tree drawn from A's point of view, A's move.

3-4.3. Searching a Partial Game Tree

For most games, the tree of possibilities is much too large to be generated fully or searched backward for an optimal move. Thus a reasonable portion of the tree is generated starting from the current position, a move is made on the basis of partial knowledge, the opponent reply found, and the procedure recursively repeated from the new position. The minimax procedure thus starts with an estimate of the tip nodes thus far generated, and assigns backed-up values to the ancestors (e.g., values in parentheses in Fig. 3-5). The value estimates for the tip nodes are generated using a static evaluation function based on heuristics.

To reduce the number of nodes that need to be examined, various pruning techniques have been devised, "Alpha-Beta" (see, e.g., Marsland, 1983) being the best known. All these techniques are based on keeping track of backed-up values so that branches that cannot lead to better solutions need not be further explored.

3-4.4. Heuristics in Game Tree Search

A *static evaluation function* is one that estimates a board position without looking at any of the positions' successors. The function is usually a linear polynomial whose variables represent various features of the position. For chess, the features of importance include remaining pieces, king safety, center control, and pawn structure.

3-4.5. Other Considerations

Alternatives to search in choosing moves include opening or end-game "book" moves, and recognizing patterns on the board and associating appropriate playing methods with each pattern. The most successful game-playing programs thus far, have made search, rather than knowledge, their main ingredient. Various combinations of more extensive use of specific game knowledge to prune less desirable paths, and increased look-ahead, have been utilized in chess in efforts to improve program success.

3-5. NONDEDUCTIVE PROBLEM-SOLVING APPROACHES

3-5.1. Backward Chaining

Backward chaining is a name given to depth-first, backward reasoning—an important search strategy. An operator is chosen that would achieve the goal if selected. If it is applicable in the initial state, it is applied and a solution has been found. If not, operators that would achieve the preconditions required for its applicability are sought and the search continues recursively until a sequence of operators are found that transform the initial state into the goal state. If the search fails, the program backtracks and a new candidate operator is selected that would achieve the goal if applied, and the process is repeated.

For problems requiring only a small amount of search, backward chaining strategies are often perfectly adequate and efficient. For larger problems, it is critical that the correct operator be chosen first almost always, because this strategy follows out a line of action fully before rejecting it, which can result in very lengthy searches.

3-5.2. Problem Reduction

A generalization of backward chaining is problem reduction. Very often to satisfy a goal, several subproblems (conjuncts) must be satisfied simultaneously. For this case of backward reasoning, applying an operator may divide the problem into a set of subproblems, each of which may be significantly simpler to solve than the original problem.

A good example of problem reduction is readying a space vehicle for launch, as indicated in Fig. 3-6. Note that we can represent the goal—spacecraft ready to launch—as a conjunction of subgoals (e.g., spacecraft fueled, all systems checked, power on). These in turn can consist either of a set of simultaneous ("AND") subgoals, or of one of several acceptable alternatives ("OR" subgoals). The AND subgoals are denoted on the graph by horizontal arcs connecting the lines leading to them.

Problem reduction often runs into difficulties without specific problem knowl-

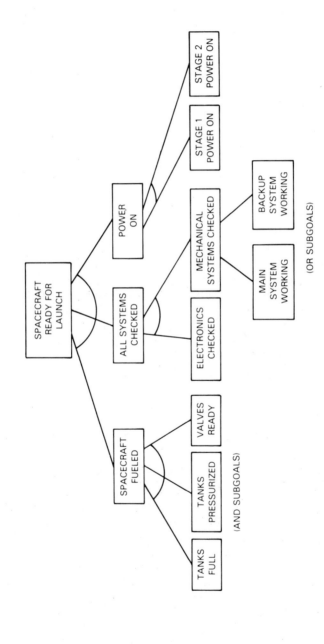

Figure 3-6 Simplified AND/OR graph for readying a spacecraft for launch.

edge, as there is otherwise no good reason to attack one interacting conjunct before another. Lack of such knowledge may lead to an extensive search for a sequence of actions that tries to achieve subgoals in an unachievable order.

3-5.3. Difference Reduction ("Means-Ends" Analysis)

Difference reduction was introduced by the General Problem Solver (GPS) program developed by Allen Newell, J. C. Shaw, and Herbert Simon, beginning in 1957. This was the first program to separate its general problem-solving method from knowledge specific to the current problem.

The means-ends analysis first determines the difference between the initial and goal states and selects the particular operator that would most reduce the difference. If this operator is applicable in the initial state, it is applied and a new intermediate state is created. The difference between this new intermediate state and the goal state is then calculated and the best operator to reduce this difference is selected. The process proceeds until a sequence of operators is determined that transforms the initial state into the goal state.

The difference reduction approach assumes that the differences between a current state and a desired state can be defined and the operators can be classified according to the kinds of differences they can reduce. If the initial and goal states differ by a small number of features, and operators are available for individually manipulating each feature, difference reduction works. However, there is no inherent way in this approach to generate the ideas necessary to plan complex solutions to difficult problems.

3-5.4. More Efficient Tactics for Problem Solving

For more efficient problem solving it is necessary to devise techniques to guide the search by making better use of initial knowledge about the problem or of the information that can be discovered or learned about the problem as the problem solver proceeds through the search.

Sacerdoti (1979) indicates that information relevant to planning that can be learned during the exploration process includes:

- Order relationships among actions
- Hierarchical links between actions at various levels of abstraction
- The purpose of the actions in the plan
- The dependence among objects (or states) being manipulated

There are two opposing ways to improve the efficiency (solution time) of a problem solver:

- Use a cheap (quickly calculable) evaluation function and explore lots of paths that might not work out, but in the process acquire information about the interrelationships of the actions and the states as an aid in efficiently guiding a subsequent search.

- Use a relatively expensive evaluation function and try hard to avoid generating states not on the eventual solution path.

The following methods are attempts to achieve more efficient problem solving through employing various ratios of exploration and evaluation.

Hierarchical planning and repair.   As in planning by human beings, one can start by devising a general plan and refine it several times into a detailed plan. The general plan can be used as a skeleton for the more detailed plan. Using this approach, generating rather complex plans can be reduced to a hierarchy of much shorter, simpler subproblems. As the detailed plans are generated, the results should be checked to see that the intended general plan is being realized. If not, various methods for patching up the failed plan can be applied.

Another approach is to observe that some aspects of a problem are significantly more important than others. By utilizing this hierarchical ranking, a problem solver can concentrate most of its efforts on the critical decisions or more important subgoals first.

Problem solving by creating and then debugging almost-right plans.   This approach deliberately oversimplifies the problem so that it can be solved more readily and then corrects the solution using special debugging techniques (associated with errors due to the simplification). An everyday example is the general tactic by which people use road maps: Find a simple way to get to the vicinity of your destination and then refine the plan from there.

Special-purpose subplanners.   This approach uses built-in subroutines to plan frequently occurring portions of a problem, such as certain moves or subgoals in robotics.

Constraint satisfaction.   This technique provides special-purpose subplanners to help ensure that the action sequences that are generated will satisfy constraints.

Relevant backtracking (dependency-directed or nonchronological backtracking).   The focus here is on sophisticated postmortem analysis gained from several attempts that failed. The problem solver then uses this information to backtrack, not to the most recent choice point but to the most relevant choice point.

Disproving.   In this approach, attempts are made to prove the impossibility of the goal, both to avoid further pursuit of an intractable problem and to employ the resultant information generated to help suggest an action sequence to achieve the goal for a feasible problem.

Pseudo-reduction.   For the difficult case where multiple goals (conjuncts) must be satisfied simultaneously, one approach is to find a plan to achieve each conjunct independently. The resultant solutions to these simpler problems are then integrated using knowledge of how plan segments can be intertwined without destroying their important effects. By avoiding premature commitments to particular orderings of subgoals, this tactic eliminates much of the backtracking typical of problem-solving systems.

Goal regression.   This tactic regresses the current goal to an earlier position in the list of goals to be satisfied. This approach can be useful in cases where conjunctive subgoals must be satisifed, but where the action that satisfies one goal tends to interfere with the satisfaction of the others.

Table 3-1 indicates where the emphasis lies in the various problem-solving techniques discussed—either in the computational effort employed in evaluating the information gained thus far from the searched region or in the effort expended in choosing the next move based only on local information.

TABLE 3-1   Primary Emphasis of Problem-Solving Tactics

Relationship	Learn and Evaluate	Choose New Move Based on Local Information
Sequencing order	Pseudo-reduction (plan generation portion) Relevant backtracking Disproving	
Hierarchy	Plan and repair	Special-purpose subplanner
Purpose of actions	Creating almost-right plans Pseudoreduction (plan repair portion)	Goal regression
Dependency among objects	Relevant backtracking	Constraint satisfaction

Source: Derived from Sacerdoti (1979, p. 15).

3-5.5. Production Systems

Production rules (PRs), such as:

If the shuttle power supply fails
  and a backup is available,
  and the cause of failure no longer exists,
Then switch to the backup.

have proved such a convenient modular way to represent knowledge that they now form the basis of most expert systems.*

*Such "if-then" rules are elaborated on in Section 4-2.

The basic automated problem-solving relationships of Fig. 3-2 can be recast as a production system as shown in Fig. 3-7. A production system consists of a knowledge base of production rules (consisting of domain facts and heuristics), a global data base (GDB) which represents the system status, and a rule interpreter (control structure) for choosing the rules to execute. In a simple production rule system, the rules are tried in order and executed if their preconditions (the "if" portion of the rules) match a pattern in the GDB.

However, in more complex systems, such as used in expert systems, a very complex control structure may be used to decide which group of PRs to examine and which to execute from the PRs in the group that match patterns in the GDB. In general, these control structures work in a repetitive cycle of the form:

1. Find the "conflict set" (the set of competing rules whose preconditions match some data in the GDB).
2. Choose a rule from those in the conflict set.
3. Execute the rule, modifying the GDB.

Production rule systems can be implemented for any of the problem-solving approaches discussed earlier. Thus we may use a *top-down* approach, employing the rules to chain backward from the goal to search for a complete supportive or causal set of rules and data (*goal-driven* control structure). Or we can use a *bottom-up* approach employing forwarding-chaining of rules to search for the goal (*event-driven* or *data-driven* control structure).

In complex systems (employing many rules) the control structure may contain meta-rules which select the relevant rules from the entire set of rules and focus attention on the relevant part of the data base. This reduces the search space to be considered. The control structure then employs further heuristics to select the most

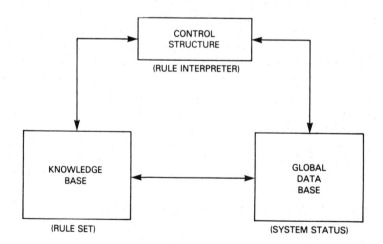

Figure 3-7 A production system.

appropriate rule from the conflicting rules that match the preconditions in the global data base. Johnson (1980, p. 7) describes this approach as follows:

> Event-driven logic operates in the forward direction, comparing the left-hand sides of the rules in the rule-set with the data in the data base. The "best" of the matching rules found is selected and fired, causing the righthand side of that rule to make some modification in the global data base. This process is repeated until a goal rule matches the data and terminates the process . . . a goal rule is one which tests whether the problem is done.

In a "pure" production rule interpreter, the generalized repeating process takes the form shown in Fig. 3-8. That idealization may be considered as a four-cycle logical process with activation, matching, conflict-resolution, and execution subcycles. For generality, the subcycle machinery in Fig. 3-8 is shown to be controlled by additional sets of higher-order rules about rules, which are called meta-rules. In practice, one often finds part or all of the meta-rule machinery of Fig. 3-8 replaced by simpler mechanisms. In practical programs many variations of this basic scheme exist because of efficiency considerations, the characteristics of the particular applications, and programmer preferences.

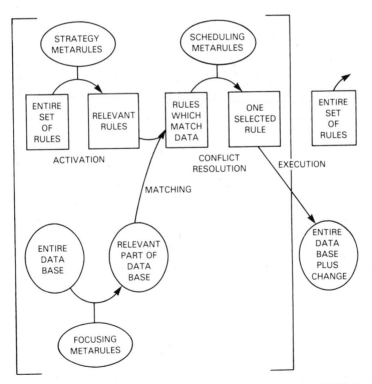

Figure 3-8 Idealized event-driven control scheme. (From Johnson, 1980) Used by permission of the author.

A simple example of an event-driven production system can be visualized for a shuttle flight in which the power supply status is observed to be out of limits in the global data base. The strategy meta-rules indicated in Fig. 3-8 then select, from the tens of thousands of rules in the knowledge base having to do with shuttle flight operations, those rules having to do with power and the use of power. Similarly, the focusing meta-rules select from the GDB the relevant part having to do with the status of the power supply, and the shuttle's and the experiments' use of power. The relevant rules are then compared with the relevant part of the GDB to determine which rules are appropriate for the current system status. The scheduling meta-rules (using priorities) then select the most appropriate rule (such as switching in the backup or turning off the less important experiments). Executing the selected rule changes the system status, and the cycle repeats.

3-6. CURRENT STATE OF THE ART

Real, complex problems tend to have the characteristic that their search space tends to expand exponentially with the number of parameters involved. This type of problem is still out of bounds for most large searches that do not have powerful heuristics to guide them. Chess has been one indicator of the state of the art in problem solving emphasizing search (although computer capability has been an equally important factor). Berliner (1981) reports that 1981 chess programs (emphasizing look-ahead) had reached an expertise of 2300 points compared with roughly 2500 points for the best human experts.

3-7. PLAYERS AND RESEARCH TRENDS

Current problem solvers emphasizing search have thus far succeeded only in solving elementary or toy problems, or very well structured problems such as games. Thus the AI community's emphasis has shifted toward expert systems (Duda, 1981) as problem solvers, where the emphasis is on knowledge rather than search.* In addition, there are trends toward distributed problem-solving systems and toward interactive problem solving systems where human beings make the major decisions and the computer program offers choices and works out the details.

In the areas of developing search and basic problem-solving techniques, Stanford University, SRI, MIT, Carnegie-Mellon University, the University of California at Los Angeles, and the University of Maryland tend to dominate.

---

*Even in chess there is beginning to be an emphasis on knowledge, as evidenced by the CHUNKER Program (Campbell and Berliner, 1983), where the incorporation of knowledge about patterns of chess positions drastically reduces search requirements in appropriate situations.

3-8. FUTURE DIRECTIONS

It is expected that within the next five years, the increased speed and capability of computers and the ability to do parallel searches could have as much effect on search performance as new search methods. However, as search usually grows exponentially with depth, heuristics to restrict the paths to be searched will also be of continuing importance. It is also expected that techniques to combine shallow and deep reasoning (e.g., nonmonitonic reasoning, causality, first principles, theorem proving) will be major contributors to limiting and guiding search.

Schank (1983) states that "search is one of the key AI problems. However ... the approaches to search have been inadequate. Searching massive amounts of information requires not efficient algorithms but representations that obviate the need for these algorithms." (Knowledge representation is the subject of the next chapter.)

REFERENCES

Barr, A., and Feigenbaum, E. A., *The Handbook of Artificial Intelligence*, Vol. 1. Los Altos, CA: W. Kaufmann, 1981.

Berliner, H. J., "An Examination of Brute Force Intelligence," *Proceedings of the Seventh International Conference on Artificial Intelligence*, Aug. 1981, pp. 581-587.

Campbell, M., and Berliner, H., "A Chess Program That Chunks," *Proceedings of the National Conference on Artificial Intelligence, AAAI-83*, Washington, DC, Aug. 22-26, 1983, pp. 49-53.

Duda, R. O., "State of Technology in Artificial Intelligence," in *Research Directions in Software Technology*, P. Wegner, (Ed.). Cambridge, MA: MIT Press, 1979, pp. 729-749.

Duda, R. O., "Knowledge-Based Expert Systems Come of Age," *Byte*, Vol. 6, No. 9, Sept. 1981, pp. 238-281.

Johnson, C. K., "Programming Methodology of Artificial Intelligence," in *Computing in Crystallography*, R. Diamond and S. R. Kenkatsan (Eds.). Bangladore, India: Indian Academy of Sciences, 1980, pp. 28.01-28.16.

Marsland, T. A., "Relative Efficiency of Alpha-Beta Implementations," *Proceedings of the Eighth International Conference on Artificial Intelligence: IJCAI-83*, Karlsruhe, W. Germany, Aug. 8-12, 1983. Los Altos, CA: W. Kaufmann, 1983, pp. 763-766.

Pearl, J. (Ed.), *Search and Heuristics*. New York: Elsevier, 1984.

Sacerdoti, E. D., "Problem Solving Tactics," *AI Magazine*, Vol. 2, No. 1, 1979, pp. 7-14.

Schank, R. C., "The Current State of AI: One Man's Opinion," *AI Magazine*, Vol. 4, No. 1, Winter-Spring 1983.

4

KNOWLEDGE REPRESENTATION

4-1. KNOWLEDGE AS THE CORE OF AI

Artificial intelligence views knowledge as the key to high-performance intelligent systems. Thus the representation and management of knowledge is a core topic in AI today. The purpose of knowledge representation (KR) is to organize required information into a form such that the AI program can readily access it for making decisions, planning, recognizing objects and situations, analyzing scenes, drawing conclusions, and other cognitive functions. Thus knowledge representation is especially central to expert systems, computational vision, and natural language understanding.

4-2. METHODS OF KNOWLEDGE REPRESENTATION

Representation schemes are classically divided into declarative and procedural ones. *Declarative* refers to representation of facts and assertions, while *procedural* refers to actions, or what to do. A further subdivision for declarative (object-oriented) schemes includes relational (semantic network) schemes and logical schemes. The principal KR schemes are discussed briefly in the following paragraphs.

4-2.1. Logical Representation Schemes

The principal method for representing a knowledge base logically is to employ *first-order predicate logic*. In this approach, a knowledge base (KB) can be viewed as a collection of logical formulas which provide a partial description of the world. Modifications to the KB result from additions or deletions of logical formulas.

An example of a logical representation is

IN (SHUTTLE,ORBIT) = shuttle is in orbit

Logical representations are easy to understand and have available sets of inference rules needed to operate upon them. (This will be discussed in more detail in Chapter 5). A drawback of logical representation is its tendency to consume large amounts of memory.

4-2.2. Semantic Networks

A semantic network is an approach to describing the properties and relations of objects, events, concepts, situations, or actions by a directed graph consisting of nodes and labeled edges (arcs connecting nodes). A simple example is given in Fig. 4-1. Because of their way of representing associations, semantic networks are very popular in AI.

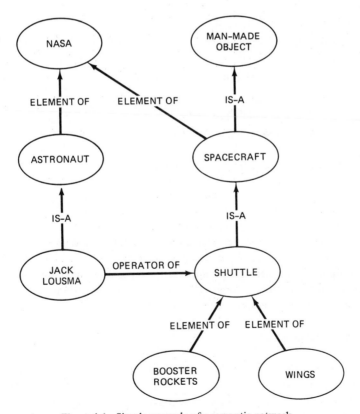

Figure 4-1 Simple example of a semantic network.

4-2.3. Procedural Representations and Production Systems

In procedural representations, knowledge about the world is contained in procedures—small programs that know how to do specific things (how to proceed in well-specified situations). Classification of procedural representation approaches are based on the choice of activation mechanisms for the procedures and the forms used for the control structures.

The two common approaches consist of procedures representing major chunks of knowledge—subroutines—and more modular procedures, such as the currently popular *production rules*. The common activation mechanism for procedures is matching the system state to the preconditions needed for the procedure to be invoked.

Production rules are characterized by a format of the type

Pattern, action

If, then

Antecedent, consequent

Situation, response

For example:

If the shuttle doors fail to close automatically when actuated, and the fault cannot be discovered,

Then disengage motors and close doors manually.

Because of their modular representation of knowledge and their easy expansion and modifiability, production rules are now probably the most popular AI knowledge representation, being chosen for most expert systems.

4-2.4. Analogical or Direct Representations

In many instances it is appropriate to use natural representations, such as an array of brightness values for an image or a further reduced "sketch map" of the scene delineations in a computer vision system. These natural representations are useful in computational vision, spatial planning, geometric reasoning, and navigation.

This form of representation has the advantages of being easy to understand, simple to update, and often allows important properties to be observed directly, so that they do not have to be inferred. However, the representation can be clumsy for some tasks and may be inappropriate when generalization is needed.

4-2.5. Property Lists

One approach to describe the state of the world is to associate with each object a property list, that is, a list of all those properties of the object pertinent to the state description. The state, and therefore the object property values, can be updated when a situation is changed.

4-2.6. Frames and Scripts

A large proportion of our day-to-day activities are concerned with stereotyped situations such as going to work, eating, shopping, and so on. Minsky (1975) conceived of *frames*, which are complex data structures for representing stereotyped objects, events, or situations. A frame has slots for objects and relations that would be appropriate to the situation. Attached to each frame is information such as:

- How to use the frame
- What to do if something unexpected happens
- Default values for slots

Frames can also include procedural as well as declarative information. Frames facilitate expectation-driven processing—reasoning based on seeking confirmation of expectations by filling in the slots. Frames organize knowledge in a way that directs attention and facilitates recall and inference.

An example of a frame is:

Airplane frame:
Type:
  range: (fighter, transport, trainer, bomber, light plane, observation)
Manufacturer:
  range: (McDonnell-Douglas, Boeing, etc.)
Empty weight:
  range: (500 to 250,000 lb)
Gross weight:
  range: (700 to 500,000 lb)
  if needed: (1.6 X empty weight)
Maximum cruising range:
  if needed: (look up in table cruising range appropriate to type and gross weight)
Number of cockpit crew:
  range: (1 to 3)
  default: 2

Scripts are framelike structures designed for representing stereotyped sequences of events such as eating at a restaurant or a newspaper report of an apartment fire.

4-2.7. Semantic Primitives

For any knowledge representation scheme, it is necessary to define an associated vocabulary. For semantic nets, there has been a real attempt to reduce the relations to a minimum number of terms (semantic primitives) that are nonoverlapping. A similar effort has emerged for natural language understanding, for which several attempts have been made to describe all the world's aspects in terms of primitives that are unique, unambiguous representations into which natural language statements can be converted for later translation into another language or for other cognitive actions.

Schank (see, e.g., Schank and Riesbeck, 1981) has developed a *conceptual dependency theory* for natural language, in an attempt to provide a representation of all actions in terms of a small number of primitives. The system relies on 11 primitive physical, instrumental, and mental *acts* (propel, grasp, speak, attend, etc.), plus several other categories or concept types. There are two basic kinds of combinations or conceptualizations. One involves an actor doing a primitive act; the other involves an object and a description of its state. Attached to each primitive act is a set of inferences that could be associated with it.

An example of a representation in conceptual dependency is:

"Armstrong flew to the moon."
Actor:	Armstrong
Action:	flew
Direction to:	the moon
From:	unknown

4-3. REPRESENTATION LANGUAGES

A number of programming languages have been designed to facilitate knowledge representation. Table 4-1 lists some of the more popular research languages. It will be observed that usually one form of knowledge representation (such as production rules or frames) is chosen as central to the language, although some (such as UNITS) provide for multiple representations.

4-4. CURRENT STATE OF THE ART

Although production rules have emerged as the dominant KR for expert systems, and semantic networks for image understanding, KR is still in a state of flux with many researchers, various representations being used, and no clear general understanding of which representations are most appropriate for which problems. As a result, KR research is one of the most active areas in AI today.

4-5. PLAYERS AND RESEARCH TRENDS

The principal researchers in KR are the universities. In the United States, these include Stanford University, Carnegie-Mellon University, the University of Pittsburgh, MIT, Yale University, the University of Maryland, the State University of

TABLE 4-1   Programming Tools Facilitating Knowledge Representation

Tools	Organization	Nature
OPS 5	Carnegie-Mellon Univ.	Programming language, built on top of LISP, designed to facilitate the use of production rules
ROSIE	Rand	General rule-based programming language that can be used to develop large knowledge bases; translates near-English into INTERLISP
UNITS	Stanford Univ.	Knowledge representation language and interactive knowledge acquisition system; the language provides both for "frame" structures and production rules
KRL	Xerox PARC	Knowledge representation language developed to explore frame-based processing
SAM	Yale Univ.	System of computer programs to analyze scripts
FRL	MIT	Frame representation language that provides a hierarchical knowledge base format consisting of frames whose slots carry comments, default values, constraints, and procedures that are activated when the value of the slot is needed
KL-ONE	Bolt, Beranek and Newman	Uniform language for representation of natural language conceptual information, based on the idea of structured inheritance networks; networks use epistemological primitives as links
NETL	Carnegie-Mellon Univ.	Comprehensive, domain-independent, knowledge-based system. It uses a parallel intersection technique for searching rapidly through large bodies of knowledge
DAWN	Digital Equipment Corp.	General programming and system description language with automated help procedures
OWL	MIT	Semantic network knowledge representation language for use in natural language question answering and for building expert systems
FRAIL	Brown Univ.	Knowledge representation that combines predicate calculus with frame representation for use in natural language understanding

New York at Buffalo, and the University of California at Berkeley. Also active are SRI; Bolt, Beranek and Newman; IBM; and Digital Equipment Corporation.

Areas of research include improved KR languages; methods to handle imprecise knowledge, intentions, and beliefs; representations of processes that consist of sequenced actions over time; representations for complex and amorphous shapes; and techniques for indexing into a large data base of models.

4-6. FUTURE DIRECTIONS

The knowledge representation field has begun to exhibit some structure—rule-based systems predominating in expert systems, but network representations also being important.* For image-understanding systems, direct representations (such as line sketches) are common, with network representations being widely employed.

In the future we will probably see increased standardization of terminology, standardized primitives, and the use of multiple types of representations in a single problem. We can also expect increased emergence of self-reflective systems that can reason about their own structure and knowledge. Also emerging will be knowledge representation systems that are appropriate for learning, generalization, and abstraction—currently difficult subjects.

KR languages are on the increase, which should help in constructing knowledge-based systems and encourage standardization of representations. Within the next five years, we can expect a clearer understanding of which representations are appropriate for which problems. We can also expect KBs to increase vastly in size, with KR techniques being developed to ease the addition of knowledge to them and the retrieval of knowledge from them.

REFERENCES

Minsky, M., "A Framework for Representing Knowledge," in *The Psychology of Computer Vision*, P. Winston (Ed.). New York: McGraw-Hill, 1975, pp. 211–277.

Schank, R. C., and Riesbeck, C. K., *Inside Computer Understanding*. Hillsdale, NJ: Lawrence Erlbaum, 1981.

*Increasingly common are "framework" systems in which the nodes in the network are frames.

5

COMPUTATIONAL LOGIC

5-1. ROLE OF COMPUTATIONAL LOGIC

It is frequently necessary to develop computer programs to deduce facts that are not explicitly represented but that are implied by other represented facts. For example, an intelligent robot may have to use logical facts about its environment to deduce when a goal state has been reached or how to reach the goal state in the first place. A data base query system may have to deduce desired information from other information in the data base. Similarly, it is often necessary to determine if a given hypothesis (theorem) follows from a given set of premises. (Such a determination is referred to as *theorem proving*.) Computational logic has been developed to address such problems.

Traditional computational logic—a computational approach to logical reasoning—is divided into two principal parts, the simpler *propositional logic* and the more complex *predicate logic*.

5-2. PROPOSITIONAL LOGIC

In logic a *proposition* is simply a statement that can be true or false. Rules used to deduce the truth (T) or falsehood (F) of new propositions from known propositions are referred to as *argument forms* or inference methods. The interesting and useful things we can do with propositions result from joining propositions together with connectives such as OR, AND, NOT, and IMPLIES to make new propositions. The symbols for these connectives are given in Fig. 5-1.

Connective	Symbol	Meaning
And	$\wedge$ or $\cap$	both
Or	$\vee$ or $\cup$	either or both
Not	$\neg$ or $\sim$	the opposite
Implies	$\supset$ or $\longrightarrow$	If the term on the left is true, then the term on the right will also be true.
Equivalent	$\equiv$	has the same truth value

Figure 5-1   Typical mathematical logic symbols.

The simplest argument form is the *conjunction*, which utilizes the connective AND. It states that if proposition p is true and proposition q is true, the conjunction "p AND q" is true. In symbolic form we have

$$p \qquad \text{(premise)}$$
$$\underline{q \qquad \text{(premise)}}$$
$$p \wedge q \qquad \text{(conclusion)}$$

which simply states that for a conjunction, the conclusion is true if both premises are true.

Deduction means obtaining solutions to problems using some systematic reasoning procedure to reach conclusions from stated premises. (In mathematical logic, deductive procedures are sometimes referred to as *formal inference*.)

One simple form of deduction can be represented as a mathematical form of argument called *modus ponens* (MP):

$$p \qquad\qquad\quad \text{(premise)}$$
$$\underline{p \text{ IMPLIES } q \quad \text{(premise)}}$$
$$q \qquad\qquad\quad \text{(conclusion)}$$

or in logical notation as

$$(p \wedge (p \rightarrow q)) \rightarrow q$$

An example of MP is

I'm feeling very sick	(premise)
When I'm feeling very sick, I must call the doctor	(premise)
I must call the doctor	(conclusion)

The conclusion is usually stated as a theorem to be proved.

Today, as an outgrowth of a general syntactic method developed by Wang (1960) at Harvard University, computer programs are available for solving problems in propositional logic.

5-3. PREDICATE LOGIC

Propositional logic is limited in that it deals only with the truth or falsehood of complete statements and does not take into account data dependency. Predicate logic remedies this situation by allowing you to deal with assertions about items in statements and allows the use of variables and functions of variables.

Propositions make assertions about items (individuals). A *predicate* is the part of the proposition that makes an assertion about the individuals and is written as

$$\overbrace{\text{Predicate (individual, individual, } \dots)}^{\text{arguments of the predicate}}$$

For example,

"The box is on the table" (proposition) is denoted as

ON (BOX, TABLE)

The predicate, together with its arguments, is a proposition. Any of the operations of propositional logic may be applied to it.

By including variables for individuals, predicate logic enables us to make statements that would be impossible in propositional logic. This can be further extended by the use of functions of variables. Finally, by use of the universal and existential quantifiers $\forall$ (for all) and $\exists$ (there exists), we arrive at first-order predicate logic (FOPL). FOPL permits rather general statements to be made, for example,

For all earth satellites, there exists a point y on the satellite that is closest to the earth:

$$\forall(x) \, \text{SATELLITE}(x) \rightarrow \exists \, (y) \, (\text{CLOSEST}(y, \text{earth}) \wedge \text{ON}(y, x))$$

Various inference rules exist for the manipulation of quantifiers, the substitution of connectives, and other syntactic operations that assist in performing logical reasoning.

5-4. LOGICAL INFERENCE

5-4.1. Theorem Proving

Logical inference—reaching conclusions using logic—is normally done by *theorem proving*. Theorem-proving approaches can be divided into resolution-based theorem proving and non-resolution-based theorem proving. Pure resolution theorem proving is syntactic in nature. Non-resolution theorem proving is more semantic in nature, emphasizing the use of heuristics and user-supplied knowledge, techniques, and procedures (see, e.g., Bledsoe, 1981) to produce more human–like theorem provers. Non-resolution-based theorem-proving systems have reached the stage where some meaningful mathematical theorems can be proved, but do not appear to be close to solving open problems in the continuum mathematical areas such as analysis.

Resolution method. The most popular method for automatic theorem proving is the resolution procedure developed by Robinson (1965). This procedure is a general automatic method for determining if a hypothesized-conclusion (theorem) follows from a given set of premises (axioms). Resolution is a syntactic method of deduction which replaces all the many argument forms of traditional logic. Resolution is a simple concept but to discuss it, a few additional definitions will be helpful.

- *Atom:* A proposition that cannot be broken down into other propositions (i.e., a proposition that is not formed from other propositions by using connectives).
- *Literal:* A atom (e.g., q) or an atom preceded by NOT (e.g., NOT q).
- *Clause:* A series of literals joined by OR, for example, (NOT p) OR q OR (NOT r). (Duplicate literals in clauses can be eliminated. This process is called *factoring.*)
- *Resolution principle:* An argument form that applies to clauses:

$$
\begin{array}{ll}
p \text{ OR } l \text{ OR } m \text{ OR} \ldots & \text{(premise)} \\
\underline{(\text{NOT } p) \text{ OR } n \text{ OR } o \text{ OR} \ldots} & \text{(premise)} \\
l \text{ OR } m \text{ OR } n \text{ OR } o \text{ OR} \ldots & \text{(conclusion)}
\end{array}
$$

[If the premises are T, then by resolution (the cancellation of contradicting literals between clauses) the conclusion is T.] Basically, resolution is the cancellation between clauses of a proposition in one clause with the negation of the same proposition in another clause.

- *Identity:* States that two propositions are equivalent, for example,

$$\text{NOT (NOT } p) = p \quad \text{(identity)}$$

- *Empty clause:* (□) indicates a contradiction:

$$
\begin{array}{l}
p \\
\underline{\text{NOT } p} \\
\square \text{ (by resolution)}
\end{array}
$$

After first putting the original premises and the conclusion into clause form using standard identities, we are ready to prove the truth of a conclusion from a set of premises using resolution. Start by negating the desired conclusion (the theorem to be proved). Then derive new clauses using unification* followed by factoring and reso-

---

Unification is the name for the procedure for carrying out instantiations. In unification we attempt to find substitutions (instantiations) for variables that will make two atoms identical.

lution, working toward deriving the empty clause. If the empty clause is derived, then (as a result of this proof by contradiction) the desired conclusion follows from the original premises. If we stop before the empty clause is derived, either the conclusion does not follow from the premises or we gave up too soon.

Unfortunately, pure resolution has been unable to handle complex problems, as the search space generated by the resolution method grows exponentially with the number of formulas used to describe a problem. Thus for complex problems, resolution derives so many clauses not relevant to reaching the final contradiction that it tends to use up the available time or memory before reaching a conclusion. Therefore, approaches are necessary that restrict the generation of irrelevant clauses.

Cohen and Feigenbaum (1982, pp. 80-81) state that "one kind of guidance that is often critical to efficient system performance is information about whether to use facts in a *forward-chaining* or *backward-chaining* manner. . . . Early theorem-proving systems used every fact both ways leading to highly redundant searches."* Another factor that can greatly affect the efficiency of the deductive reasoning is the way in which a body of knowledge is formalized. "That is, logically equivalent formalizations can have radically different behavior when used with standard deduction techniques."

The current resolution theorem-proving branch of computational logic is in large part an outgrowth of the earlier pure resolution theorem proving, with additional techniques and modifications added to attempt to restrain combinatorial explosions. With restrictions on resolution clause generation, theorem-proving approaches can be made sufficiently efficient to be used in practical problems. An outstanding example of this is the AURA (AUtomated Reasoning Assistant) theorem-proving system (Wos, 1983), which has been applied successfully to real (though limited) applications in mathematics, formal logic, program verification, logic circuit design, chemical synthesis, data base inquiry, and robotics.

Three techniques (used in the AURA system) that have had a major impact on making theorem provers practical are:

1. *Demodulation:* employing rewrite rules to simplify or canonicalize the expressions to achieve a normalized form
2. *Subsumption:* a technique that recognizes and discards many equivalent or weaker rules or facts than those that have already been generated
3. *Strategy rules:* ordering strategies that direct the system as to what to do next

These three powerful techniques in AURA are domain independent (although the strategy rules have provision for weighting so that the user can assign priorities to concepts).

Other strategies have been important for further reducing the expressions that are generated or retained during the proof process. One class is restriction strategies, which provide guidance as to which operations can be skipped. For example, there is the *set of support* strategy that discourages looking at facts that do not have support (e.g., general information used alone, unsupported by other facts). There are indications that there remain many important domain-independent inference rules yet to be discovered.

Nonresolution theorem proving.   Cohen and Feigenbaum (1982, p. 94) observe that "in *non-resolution* or *natural deduction* theorem-proving systems, a proof is derived in a goal-directed manner that is natural for humans using the theorem prover. Natural-deduction systems represent proofs in a way that maintains a distinction between goals and antecedents, and they use inference rules that mimic the reasoning of human theorem-proving." They also tend to use domain-specific heuristics that help guide the search and many proof rules to reduce goals to subgoals. The result is much more complex than the simpler, but less effective basic resolution procedure.

Although requiring help from the programmer, the nonresolution Boyer and Moore (1979) theorem prover is one of the most powerful theorem provers available.

5-4.2. Logic Programming

It was realized in the early 1970s that logic representations could also function in a procedural mode by using the technique of unification to search for instantiations that would satisfy stated goals. This has led to the PROLOG programming language (discussed in Appendix C).

As the manner in which the representations are written and the order (e.g., top to bottom, and left to right) chosen for the execution of the logic statements can have an important influence on the efficiency and effectiveness of executing the program, such representational and ordering choices can be thought of as a form of programming, hence the name *logic programming*. PROLOG, and logic programming in general, has become very popular in the last few years.

5-4.3. Nonmonotonic Logic

One of the popular issues in AI problem solving (see, e.g., McCarthy, 1981) has been concerned with how to handle lines of reasoning and conclusions that may have to be retracted when new information is received. For example, it is usually reasonable to conclude that if a creature is a bird, then it can fly. However, if it is later learned that the bird is a penguin or is dead, the conclusion must be reconsidered. Etherington and Reiter's (1983) work on providing a formal semantics for networks of inheritance hierarchies with exceptions appears particularly promising.

5-4.4. Multivalued and Fuzzy Logics

Conventional logics deal with the truth or falsity of statements. However, this binary approach is often inadequate for situations in which degrees of certainty are involved, as, for example, in medical diagnosis. Thus work in multivalued and fuzzy logics has been undertaken to attempt to address this problem.

5-5. COMMONSENSE REASONING

Commonsense is low-level reasoning based on a vast amount of experiential knowledge. An example is reasoning about falling objects, based on experience rather than on Newton's laws. The same sort of reasoning tells us what is the appropriate thing to do in everyday social situations. Although it is a simple matter for human beings, it is very difficult to achieve in present AI systems with current techniques. More generally, commonsense reasoning is everyday non-formalized reasoning. In this context, Nilsson (1980, p. 154) argues that

> many common sense reasoning tasks that one would not ordinarily formalize can, in fact, be handled by predicate calculus theorem-proving systems. The general strategy is to represent specialized knowledge about the domain as predicate calculus expressions and to represent the problem or query as a theorem to be proved.

Nilsson (1980, p. 423) also observes that "much common sense reasoning (and even technical reasoning) is inexact in the sense that the conclusions and the facts and rules on which it is based are only approximately true. Yet people are able to use uncertain facts and rules to arrive at useful conclusions about everyday subjects such as medicine. A basic characteristic of such approximate reasoning seems to be that a conclusion carries more conviction if it is independently supported by two or more separate arguments." Several of the AI expert systems, such as PROSPECTOR, make use of this approach.

5-6. FUTURE DIRECTIONS

Moore (1982) argues that a number of important features of commonsense reasoning involving incomplete knowledge of a problem situation can be implemented only within a logical framework. Thus logic-based systems will continue to be an important element of AI.

The advent of powerful resolution-based theorem-proving systems (such as AURA)—utilizing both domain-independent and domain-dependent inference rules to constrain combinatorial explosions—has resulted in opening up practical applications for such systems. However, much research remains to be done to discover more effective strategies, to devise methods for linking rules together to take larger reasoning steps, to explore parallel processing approaches, to build user-friendly interfaces, and to develop more rapid and improved knowledge representation techniques.

An important trend is the incorporation of nonresolution theorem-proving techniques into resolution-based theorem provers to provide a powerful single theorem-proving environment. The advent of portable theorem-proving systems opens up the opportunity for much increased experimentation, which should be instrumental in rapidly advancing the field. Wos (1983) predicts that as a result, automated reasoning systems with the capability for being used in a wide variety of real applications will be commonplace within five years.

In general, as theorem-proving techniques mature, they can be expected to be employed in AI applications such as deductive-like programming languages (as exemplified by PROLOG), tools for the verification of software programs, examination of the integrity of knowledge bases for expert systems, program synthesis (automatic programming), and the solving of combinatorial problems such as those found in circuit analysis.

As expert systems technology pushes forward toward employing causality and structure (in addition to empirical association rules), deeper levels of reasoning will be required. It is anticipated that advanced theorem-proving systems will play an important role in this arena.

PROLOG, the rapidly proliferating language for logic programming, has been earmarked for the Japanese Fifth-Generation Computer project. The powerful inference rules (such as the set of support strategy) used in advanced theorem provers are now being considered for use with PROLOG. These, coupled with domain-specific control strategies and making provisions for taking advantage of many of the features of LISP (as in LOGLISP, Robinson and Sibert, 1981, and in MRS, Geneserth, 1984), may well make a hybridized PROLOG the dominant AI language within the next decade. The basic logical reasoning problems (and those of nonmonotonic reasoning and reasoning in the presence of uncertainty) are beginning to succumb to some of the recent research. We can thus conclude that computational logic, which earlier appeared doomed by the combinatorics generated by the pure resolution approach, has become revitalized with new representational approaches, inference rules, domain heuristics, and advanced computers and will play an increasingly important role in future AI applications.

REFERENCES

Bledsoe, W. W., "Non-resolution Theorem Proving," in *Readings in Artificial Intelligence*, B. L. Webber and N. J. Nilsson (Eds.). Palo Alto, CA: Tioga, 1981, pp. 91–108.

Boyer, R., and Moore, J., *Computational Logic*. New York: Academic Press, 1979.

Cohen, P. R., and Feigenbaum, E. A., *The Handbook of Artificial Intelligence*, Vol. 3, Los Altos, CA: W. Kaufmann, 1982.

Etherington, D. W., and Reiter, R., "On Inheritance Hierarchies with Exceptions," *AAAI-83*, pp. 104–108.

Geneserth, M. R., "Partial Programs," Department of Computer Science Memo HPP84-1, Stanford University, Stanford, CA, Jan. 1984.

McCarthy, J., "Circumscription—A Form of Non-monotonic Reasoning," in *Readings in Artificial Intelligence*, R. L. Webber and N. J. Nilsson (Eds.). Palo Alto, CA: Tioga, 1981, pp. 466–472.

Moore, R. C., "The Role of Logic in Knowledge Representation and Common Sense Reasoning," *Proceedings of the National Conference on Artificial Intelligence*, Pittsburgh, PA:, Aug. 1982, Los Altos, CA: W. Kaufmann, 1982, pp. 428–433.

Nilsson, N. J., *Principles of Artificial Intelligence*. Palo Alto, CA: Tioga, 1980.

Robinson, J. A., "Machine-Oriented Logic Based on the Resolution Principle," *Journal of the Association for Computing Machinery*, Vol. 12, 1965, pp. 33–41.

Robinson, J. A., and Sibert, E. E., Logic Programming in LISP, RADC-TR-80-379, Vol. 1, Syracuse Univ., School of Computer and Information Science, Syracuse, NY, Jan. 1981.

Wang, H., "Toward Mechanical Mathematics," *IBM Journal of Research and Development*, Vol. 4, 1960, pp. 2–22.

Wos, L., "Automated Reasoning: Real Uses and Potential Uses," *IJCAI-83*, pp. 867–876.

6

KNOWLEDGE ENGINEERING

AND EXPERT SYSTEMS

6-1. INTRODUCTION

Expert systems probably constitute the "hottest" topic in artificial intelligence today. Prior to the last decade, in trying to find solutions to problems, AI researchers tended to rely on non-knowledge-guided search techniques or computational logic. These techniques were successfully used to solve elementary problems or very well structured problems such as games. However, realistic problems exhibit the characteristic that their search space expands exponentially with the number of parameters involved. For such problems, these older techniques have generally proved to be inadequate and a new approach was needed. This new approach emphasized knowledge rather than search and has led to the field of knowledge engineering and expert systems. The resultant expert systems technology, limited to academic laboratories in the 1970s, is now becoming cost-effective and is beginning to enter into commercial applications.

6-2. NATURE OF KNOWLEDGE-BASED EXPERT SYSTEMS

Feigenbaum, a pioneer in expert systems, (1982, p. 1) states:

> An "expert system" is an intelligent computer program that uses knowledge and inference procedures to solve problems that are difficult enough to require significant human expertise for their solution. The knowledge necessary to perform at such a level, plus the inference procedures used, can be thought of as a model of the expertise of the best practitioners of the field.

The knowledge of an expert system consists of facts and heuristics. The "facts" constitute a body of information that is widely shared, publicly available, and generally agreed upon by experts in a field. The "heuristics" are mostly private, little-discussed rules of good judgement (rules of plausible reasoning, rules of good guessing) that characterize expert-level decision making in the field. The performance level of an expert system is primarily a function of the size and quality of the knowledge base that it possesses.

It has become fashionable today to characterize any large, complex AI system that uses large bodies of domain knowledge as an expert system. Thus nearly all AI applications to real-world problems can be considered in this category, although the designation "knowledge-based systems" is more appropriate. The uses of expert systems are virtually limitless. They can be used to diagnose, monitor, analyze, interpret, consult, plan, design, instruct, explain, learn, and conceptualize.

6-3. THE STRUCTURE OF EXPERT SYSTEMS

An *expert system* consists of:

1. A knowledge base (or knowledge source) of domain facts and heuristics associated with the problem
2. An inference procedure (or control structure) for utilizing the knowledge base in the solution of the problem
3. A working memory—*global data base*—for keeping track of the problem status, the input data for the particular problem, and the relevant history of what has thus far been done

A human "domain expert" usually collaborates to help develop the knowledge base. Once the system has been developed, in addition to solving problems, it can be used to help instruct others in developing their own expertise.

It is desirable, although not yet common, to have a user-friendly natural language interface to facilitate the use of the system in all three modes: development, problem solving, instruction. Menu-driven and graphics interfaces are currently often employed. In some sophisticated systems, an explanation module is also included, allowing the user to challenge and examine the reasoning process underlying the system's answers. Figure 6-1 is a diagram of an idealized expert system. When the domain knowledge is stored as production rules, the knowledge base is often referred to as the *rule base* and the inference engine as the *rule interpreter*.

An expert system differs from more conventional computer programs in several important respects. Duda (1981, p. 242) observes that in an expert system "there is a clear separation of general knowledge about the problem (the rules forming a knowledge base) from information about the current problem (the input data) and the methods for applying the general knowledge to the problem (the rule inter-

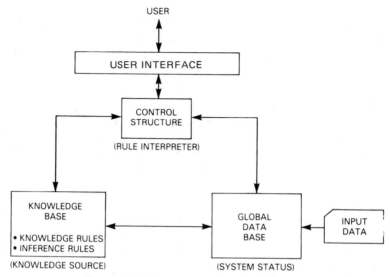

Figure 6-1 Basic structure of an expert system.

preter)." In a conventional computer program, knowledge pertinent to the problem and methods for utilizing this knowledge are all intermixed, so that it is difficult to modify the program. In an expert system, "the program itself is only an interpreter (or general reasoning mechanism) and (ideally) the system can be changed by simply adding or subtracting rules in the knowledge base."

6-3.1. The Knowledge Base

The most popular approach to representing the domain knowledge (both facts and heuristics) needed for an expert system is by production rules (also referred to as *situation-action rules* or *if-then rules*).* Thus, often a knowledge base is made up mostly of rules which are invoked by pattern matching with features of the task environment as they currently appear in the global data base.

6-3.2. The Control Structure

In an expert system a problem-solving paradigm must be chosen to organize and control the steps taken to solve the problem. A common but powerful approach involves the chaining of if-then rules to form a line of reasoning. The rules are actuated by patterns (which, depending on the strategy, match either the IF or the THEN side of the rules) in the global data base. The application of the rule changes

---

*Not all expert systems are rule-based. The network-based expert systems MACSYMA, Internist/Caduceus, Digitalis Therapy Advisor, HARPY, and Prospector are examples which are not. Buchanan and Duda (1982) state that the basic requirements in the choice of an expert system knowledge representation scheme are extendibility, simplicity, and explicitness. Thus rule-based systems are particularly attractive.

the system status and therefore the data base, enabling some rules and disabling others. The rule interpreter uses a control strategy for finding the enabled rules and for deciding which of the enabled rules to apply. The basic control strategies used may be top-down (goal driven), bottom-up (data driven), or a combination of the two that uses a relaxation-like convergence process* to join these opposite lines of reasoning together at some intermediate point to yield a problem solution. However, virtually all the heuristic search and problem-solving techniques that the AI community has devised have appeared in various expert systems.

6-3.3. Architecture of Expert Systems

One way to classify expert systems is by function (e.g., diagnosis, planning, etc.). However, examination of existing expert systems indicates that there is little commonality in detailed system architecture that can be detected from this classification. A more fruitful approach appears to be to look at problem complexity and problem structure and deduce what data and control structures might be appropriate to handle these factors.

The knowledge engineering community has evolved a number of techniques [presented in the excellent tutorial by Stefik et al. (1982) and summarized in Gevarter (1982)] which can be utilized in devising suitable expert system architectures,

The use of these techniques in four existing expert systems is illustrated in Tables 6-1A through 6-1D, which outline the basic approach taken by each of these expert systems and show how the approach translates into key elements of the knowledge base, global data base, and control structure. An indication of the basic control structures of the systems in Tables 6-1A through 6-1D, and some of the other well-known expert systems, is given in Table 6-2.

Table 6-2 represents expert system control structures in terms of the search direction, the control techniques utilized, and the search space transformations employed. The approaches used in the various expert systems are different implementations of two basic ideas for overcoming the combinatorial explosion associated with search in real complex problems. These two ideas are:

1. Find ways to search a space efficiently (as discussed in Section 3-3).
2. Find ways to transform a large search space into smaller, manageable chunks that can be searched efficiently.

It will be observed from Table 6-2 that there is little architectural commonality based either on function or on domain of expertise. Instead, expert system design may best be considered as an art form, like custom home architecture, in which the chosen design can be implemented from the collection of available AI techniques in heuristic search and problem solving.

---

*A relaxation process is a problem-solving approach which iteratively assigns values to mutually constrained objects in such a manner that a consistent set of values are achieved (see, e.g., the relaxation vision example in Chapter 8).

TABLE 6-1A Characteristics of Example Expert Systems: DENDRAL

System:	DENDRAL
Institution:	Stanford University
Authors:	Feigenbaum et al., (1971)
Function:	Data interpretation

		Key Elements of:		
Purpose	Approach	Knowledge Base	Global Data Base	Control Structure
To generate plausible structural representations of organic molecules from mass spectrogram data	Derive constraints from the data. Generate candidate structures. Predict mass spectrographs for candidates. Compare with data.	Rules for deriving constraints on molecular structure from experimental data Procedure for generating candidate structures to satisfy constraints Rules for predicting spectrographs from structures	Mass spectrogram data Constraints Candidate structures	Forward chaining Plan, generate, and test

TABLE 6-1B Characteristics of Example Expert Systems: AM

System: AM
Institution: Stanford University
Author: Lenat (1976)
Function: Concept formation

Purpose	Approach	Knowledge Base	Key Elements of: Global Data Base	Control Structure
To discover mathematical concepts	Start with elementary ideas in set theory. Search a space of possible conjectures that can be generated from these elementary ideas. Choose the most interesting conjectures and pursue that line of reasoning.	Elementary ideas in finite set theory Heuristics for generating new mathematical concepts by modifying and combining elementary ideas Heuristics of "interesting-ness" for discarding bad ideas	Plausible candidate concepts	Plan, generate, and test

TABLE 6-1C Characteristics of Example Expert Systems: R1

System: R1
Institution: Carnegie-Mellon University
Author: McDermott (1982)
Function: Design

Purpose	Approach	Knowledge Base	Global Data Base	Control Structure
			Key Elements of:	
To configure VAX computer systems (from a customer's order of components)	Break the problem up into the following ordered subtasks: 1. Correct mistakes in order. 2. Put components into CPU cabinets. 3. Put boxes into unibus cabinets and put components in boxes. 4. Put panels in unibus cabinets. 5. Lay out system on floor. 6. Do the cabling. Solve each subtask and move on to the next one in the fixed order.	Properties of (roughly 500) VAX components Rules for determining when to move to next subtask based on system state Rules for carrying out subtasks (to extend partial configuration) (Approximately 1200 rules total)	Customer order Current task Partial configuration (system state)	"MATCH"[a] (data driven) (no backtracking)

[a]MATCH selects the rule whose preconditions match and is most specialized to the current state of the configuration.

TABLE 6-1D Characteristics of Example Expert Systems: MYCIN

System: MYCIN
Institution: Stanford University
Author: Shortliffe (1976)
Function: Diagnosis

Purpose	Approach	Knowledge Base	Global Data Base	Control Structure
			Key Elements of:	
To diagnose bacterial infections and make recommendations for antibiotic therapy	Represent expert judgmental reasoning as condition-conclusion rules together with the expert's "certainty" estimate for each rule. Chain backward from hypothesized diagnoses to see if the evidence supports it. Exhaustively evaluate all hypotheses. Match treatments to all diagnoses which have high certainty values.	Rules linking patient data to infection hypotheses Rules for combining certainty factors Rules for treatment	Patient history and diagnostic tests Current hypothesis Status Conclusions reached thus far, and rule numbers justifying them	Backward chaining through the rules Exhaustive search

TABLE 6-2 Control Structures of Some Well-Known Expert Systems[a]

System	Function	Domain	Search Direction				Control								Search Space Transformations				
			Forward	Backward	Forward and Backward	Event Driven	Exhaustive Search	Generate and Test	Guessing	Relevant Backtracking	Least Commitment	Multilines of Reasoning	Network Editor	Beam Search	Multiple Models	Break into Subproblems	Hierarchical Refinement	Hierarchical Resolution	Meta-Rules
MYCIN	Diagnosis	Medicine	X	X			X												
DENDRAL	Data interpretation	Chemistry						X		X									
EL	Analysis	Electrical circuits	X						X	X									
GUIDON	Computer-aided instruction	Medicine				X							X						
KAS	Knowledge acquisition	Geology	X																
META-DENDRAL	Learning	Chemistry	X					X											
AM	Concept formation	Math	X				X	X											
VM	Monitoring	Medicine				X	X	X											
GA1	Data interpretation	Chemistry	X				X												
R1	Design	Computers	X				X	X											
ABSTRIPS	Planning	Robots		X												X			
NOAH	Planning	Robots		X						X	X				X	X			
MOLGEN	Planning	Genetics			X				X	X	X					X			X
SYN	Design	Electrical circuits	X												X				
Hearsay II	Signal Interpretation	Speech understanding			X						X	X						X	
HARPY	Signal interpretation	Speech understanding	X									X							
Crysalis	Data interpretation	Crystallography				X		X											X

[a] Additional information on these systems is given in Gevarter (1982).

In addition to the techniques indicated in Table 6-2, also emerging are distributed knowledge and problem-solving approaches exemplified by the **MDX** expert system (Chandrasekaran, 1983) and the object-oriented programming language, LOOPS (Stefik et al., 1983).

6-4. CONSTRUCTING AN EXPERT SYSTEM

Duda (1981, p. 262) states that to construct a successful expert system, the following prerequisites must be met:

- There must be at least one human expert acknowledged to perform the task well.
- The primary source of the expert's exceptional performance must be special knowledge, judgment, and experience.
- The expert must be able to explain the special knowledge and experience and the methods used to apply them to particular problems.
- The task must have a well-bounded domain of application.

Using present techniques and programming tools, the effort required to develop a major expert system appears to be converging toward five person-years, with most endeavors employing two to five people in the construction.

6-5. CURRENT STATE OF THE ART

Buchanan (1981, pp. 6-7) indicates that the current state of the art in expert systems is characterized by the following:

- *Narrow domain of expertise:* Because of the difficulty in building and maintaining a large knowledge base, the typical domain of expertise is narrow. The principal exception is Internist (Miller et al., 1982; see Fig. 6-2), for which the knowledge base covers 500 disease diagnoses. However, this broad coverage is achieved by using a relatively shallow set of relationships between diseases and associated symptoms. (Internist is being replaced by Caduceus, which uses causal relationships to help diagnose simultaneous unrelated diseases.)
- *Limited knowledge representation languages for facts and relations.*
- *Relatively inflexible and stylized input/output languages.*
- *Stylized and limited explanations by the systems.*
- *Laborious construction:* At present, it requires a knowledge engineer to work with a human expert to extract and structure the information to build the knowledge base. However, once the basic system has been built, in a few cases it has been possible to write knowledge acquisition systems to help extend

Internist-1: broad but shallow

Researchers at the University of Pittsburgh, Pa., have spent 15 person-years developing Internist-1, an experimental computer-based diagnostic consultant for general internal medicine. The program, based on the expertise of internal-medicine specialist Jack Myers, can deal with about 500 diseases, and it is also able to diagnose multiple simultaneous diseases, in contrast with systems that attempt to ascribe all symptoms to a single ailment. It thus covers about three quarters of major known diseases.

For each disease, the program develops a profile by listing symptoms and their associated evoking strength, a measure of how specific each symptom is to a certain disease. It also notes the frequency of each symptom—how often the symptom will actually appear in patients who have the associated disease. The data base is then inverted, so that for given symptoms the associated diseases are indicated.

To use Internist-1, a physician enters patient data [see illustration], and, as each manifestation is entered, the program retrieves the diseases associated with it. Each disease hypothesis is scored, based on the evoking-strength and frequencies associated with the evidence from the manifestations entered. The resulting disease hypotheses are ranked by score, and Internist-1 then partitions the top hypotheses into groups of competitors.

Internist-1: Enter case name
[The physician enters the case name.]

Internist-1: Initial positive manifestations

Physician: + Age gtr than 55
+ Arthritis HX
+ Depression HX
+ Sex female
+ Chest pain lateral sharp
+ Liver enlarged moderate
+ Feces guiaiac test positive
∴

Internist-1: Initial negative findings

Physician: − Alcoholism chronic HX
− Diarrhea chronic
− Fever
− Uric acid blood increased
∴

Internist-1: Disregarding: Depression HX, weight increase recent HX, chest pain lateral exacerbation with breathing, chest pain lateral sharp.
∴
Considering: Age gtr than 55, sex female, urine dark HX, anorexia, feces light colored, jaundice, liver enlarged moderate.
∴
[Focus of Internist-1's attention is on the problem area formed by the competing top scoring disease hypotheses and their associated manifestations.]

Internist-1: Please enter findings of lipid(s) blood
[Seeking additional generic information to assist in selecting among top hypothesis.]

Physician: Go

[Physician has no additional information to enter in response.]

Internist-1: Cholesterol blood decreased? [Asks the single specific item that prompted the generic question that the physician couldn't answer.]

Physician: N/A [not applicable]

...

[The interaction continues until Internist-1 reaches a diagnosis.]

Internist-1: Conclude: Hepatitis chronic active [Internist-1 now recycles to the next group of competing hypotheses and continues until all symptoms are explained.]

Internist-1, an expert system capable of dealing with more than 500 diseases, interacts with physicians by means of a specialized vocabulary. It focuses on possible disease symptoms, considering both manifestations that are present and those that are absent.

(Two diseases are considered competitors if, taken together, they explain no more of the manifestations than either alone.) Internist-1 then seeks more data from the physician to assist in selecting one hypothesis from the diseases in the various competing groups.

When a diagnosis is chosen from the group, all observed manifestations explained by that diagnosis are removed from further consideration. The program then recycles to try to explain the remaining manifestations.

Thorough trials indicate that Internist-1 is about as good at diagnosis as the average clinician; nevertheless, it has real deficiencies. These include:

• Inability to attribute findings to their proper cause.

• Inability to synthesize a general overview in a complicated multisystem problem.

• Inability to recognize subcomponents of an illness.

• Shallow explanation capability.

• Inadequate representation of causality.

• Exhaustive listing of disease profiles in terms of manifestations, rather than in terms of intermediate states.

• Inability to allow for interdependency of manifestations.

• Inability to reason anatomically or temporally.

• Inability to recognize severity.

As a result, new programming approaches have been developed for complex reasoning processes, so that Caduceus, the successor to Internist-1, can synthesize a broad overview of a patient's condition incorporating evidence from causal relationships. Caduceus reasons by means of a model of the human body and its workings; thus, it can make inferences based on interdependencies of different organs. It also incorporates time-based reasoning abilities, so that it can draw inferences from data about the progression of disease symptoms as well as from nonchronological lists. It also has refined control structures that allow it to deal with symptoms that may be due to a number of different underlying diseases.

This move from purely empirical associations toward a system based on a model of functional relationships in the problem under consideration is typical of the current evolution of expert systems. System builders have found that explicit representation of such knowledge can improve the abilities of an expert system and also allow its empirical knowledge to be used for other purposes, such as teaching. —W.B.G.

Figure 6-2 Internist-1. Reprinted by permission from *Spectrum*, August 1983. © 1983 IEEE.

the knowledge base by direct interaction with a human expert, without the aid of a knowledge engineer.*

- *Single expert as a "knowledge czar":* We are currently limited to our ability to maintain consistency among overlapping items in the knowledge base. Therefore, although it is desirable for several experts to contribute, one expert must maintain control to ensure the quality of the data base.

- *Fragile behavior:* In addition, most systems exhibit fragile behavior at the boundaries of their capabilities. Thus, even some of the best systems come up with wrong answers for problems just outside their domain of coverage. Even within their domain, systems can be misled by complex or unusual cases, or for cases for which they do not yet have the needed knowledge or for which even the human experts have difficulty.

- *Requires knowledge engineer to operate:* Another limitation is that for most current systems only their builders or other knowledge engineers can successfully operate them—a friendly interface not having yet been constructed.

Nevertheless, Davis (1982) observes that there have been notable successes. A methodology has been developed for explicating informal knowledge. Representing and using empirical associations, five systems have been routinely solving difficult problems—DENDRAL, MACSYMA, MOLGEN, R1, and PUFF—and are in regular use. The first three all have serious users who are only loosely coupled to the system designers. DENDRAL, which analyzes chemical instrument data to determine the underlying molecular structure, has been the most widely used program (see Lindsay et al., 1980). R1, which is used to configure VAX computer systems, has been reported to be saving Digital Equipment Corporation $20 million per year, and is now being followed up with XCON. In addition, as indicated by Table 6-3, hundreds of systems have been constructed and are being experimented with, and a growing number of additional proprietary expert systems have been constructed and are in use by individual companies.

6-6. PLAYERS, SYSTEMS, AND RESEARCH TRENDS

Table 6-3 is a list, classified by function and domain of use, of many of the existing major expert systems. It will be observed that there is a predominance of systems in the medical and chemistry domains, following from the pioneering efforts at Stanford University. From the list it is also apparent that Stanford University dominates in a number of systems, followed by MIT; Carnegie-Mellon University; Bolt, Beranek and Newman; and SRI, with several dozen scattered efforts elsewhere.

The list indicates that thus far, the major areas of expert systems development have been in diagnosis, data analysis and interpretation, planning, computer-aided instruction, analysis, and automatic programming. However, the list also indicates

---

*Techniques to automate the initial knowledge acquisition process are now beginning to appear, as evidenced by the ETS system (Boose 1984).

that a number of pioneering expert systems already exist in quite a number of other functional areas. In addition, a substantial effort is under way to build expert systems as tools for constructing expert systems, a number of which, such as KEE (IntelliCorp), ARBY (Smart Systems Technology), ART (Inference Corp.), S.1 (Teknowledge), and Expert-Ease (Intelligent Terminals Ltd.), have already been marketed commercially.

6-7. FUTURE DIRECTIONS

Figure 6-3 lists some of the expert systems applications currently under development. There appear to be few domain or functional limitations in the ultimate use of expert systems. However, the nature of expert systems is changing. The limitations of rule-based systems are becoming apparent. Not all knowledge can be readily structured in the form of *empirical associations*. Empirical associations tend to hide causal relations (present only implicitly in such associations). Empirical associations are also inappropriate for highlighting structure and function.

Thus the newer expert systems are adding deep knowledge having to do with causality and structure. These systems will be less fragile, thereby holding the promise of yielding correct answers often enough to be considered for use in autonomous systems, not just as intelligent assistants.

The other change is a trend toward an increasing number of non-rule-based systems. These systems, utilizing semantic networks, frames, and other knowledge representations, are often better suited for causal modeling and representing structure. They also tend to simplify the reasoning required by providing knowledge representations more appropriate for the specific problem domain.

Figure 6-4 (based largely on the Hayes-Roth *IJCAI-81* Expert System Tutorial and on Feigenbaum, 1982) indicates some of the future opportunities for expert systems. Again no limitation is apparent. It appears that expert systems will eventually find use in most endeavors which require symbolic reasoning with detailed professional knowledge—which includes much of the world's work. In the process, there will be exposure and refinement of the previously private knowledge in the various fields of applications.

- Medical diagnosis and prescription
- Medical knowledge automation
- Chemical data interpretation
- Chemical and biological synthesis
- Mineral and oil exploration
- Planning/scheduling
- Signal interpretation
- Signal fusion—situation interpretation from multiple sensors
- Military threat assessment
- Tactical targeting
- Space defense

- Air traffic control
- Circuit diagnosis
- VLSI design
- Equipment fault diagnosis
- Computer configuration selection
- Speech understanding
- Intelligent Computer-Aided Instruction
- Automatic Programming
- Intelligent knowledge base access and management
- Tools for building expert systems

Figure 6-3 Expert system applications now under development.

TABLE 6-3 Existing Expert Systems by Function

Function	Domain	System[a]	Institution
Diagnosis	Medicine	PIP	MIT
	Medicine	CASNET	Rutgers Univ.
	Medicine	Internist/Caduceus	Univ. of Pittsburgh
	Medicine	MYCIN	Stanford Univ.
	Medicine	PUFF	Stanford Univ.
	Medicine	MDX	Ohio State Univ.
	Computer faults	DART	Stanford Univ./IBM
	Computer faults	IDT	Digital Equipment Corp.
	Nuclear reactor accidents	REACTOR	EG&G Idaho, Inc.
	Telephone lines	WAVE	Bell Labs
Data analysis and interpretation	Oil well logs	Dip-Meter Advisor	MIT/Schlumberger
	Chemistry	DENDRAL	Stanford Univ.
	Chemistry	GA1	Stanford Univ.
	Geology	Prospector	SRI
	Protein crystallography	Crysalis	Stanford Univ.
	Determination of causal relationships in medicine	RX	Stanford Univ.
	Determination of causal relationships in medicine	ABEL	MIT
	Oil well logs	ELAS	AMOCO
Analysis	Electrical circuits	EL	MIT
	Symbolic mathematics	MACSYMA	MIT
	Mechanics problems	MECHO	Univ. of Edinburgh
	Naval task force threat analysis	TECH	Rand/Naval Ocean Systems Center, San Diego
	Earthquake damage assessment for structures	SPERIL	Purdue Univ.
	Digital circuits	CRITTER	Rutgers Univ.
Design	Computer system configurations	R1/XCON	Carnegie-Mellon Univ./Digital Equipment Corp.
	Circuit synthesis	SYN	MIT
	Chemical synthesis	SYNCHEM	State University of New York, Stonybrook

Task	System	Application	Developer
Planning	SECHS	Chemical synthesis	Univ. of California, Santa Cruz
	NOAH	Robotics	SRI
	ABSTRIPS	Robotics	SRI
	DEVISER	Planetary flybys	JPL
	OP-PLANNER	Errand planning	Rand
	MOLGEN	Molecular genetics	Stanford Univ.
	KNOBS	Mission planning	MITRE
	ISIS-II	Job shop scheduling	Carnegie-Mellon Univ.
	SPEX	Design of molecular genetics experiments	Stanford Univ.
	HODGKINS	Medical diagnosis	MIT
	AIRPLAN	Naval aircraft ops	Carnegie-Mellon Univ.
	TATR	Tactical targeting	Rand
Learning from experience	META-DENDRAL	Chemistry	Stanford Univ.
	EURISKO	Heuristics	Stanford Univ.
Concept formation	AM	Mathematics	Carnegie-Mellon Univ.
Signal interpretation	Hearsay II	Speech understanding	Carnegie-Mellon Univ.
	HARPY	Speech understanding	Carnegie-Mellon Univ.
	SU/X	Machine acoustics	Stanford Univ.
	HASP	Ocean surveillance	System Controls, Inc.
	STAMMER-2	Sensors on board naval vessels	Naval Ocean Systems Center, San Diego/SDC
Monitoring	ALVEN	Medicine—left-ventrical performance	Univ. of Toronto
	ANALYST	Military situation determination	MITRE
Use advisor	VM	Patient respiration	Stanford Univ.
	SACON	Structural analysis computer program	Stanford Univ.
Computer-aided instruction	SOPHIE	Electronic troubleshooting	Bolt, Beranek and Newman
	GUIDON	Medical diagnosis	Stanford Univ.
	EXCHECK	Mathematics	Stanford Univ.
	STEAMER	Steam propulsion plant operation	Bolt, Beranek and Newman
	BUGGY	Diagnostic skills	Bolt, Beranek and Newman
	WHY	Causes of rainfall	Bolt, Beranek and Newman
	WEST	Coaching of a game	Bolt, Beranek and Newman
	WUMPUS	Coaching of a game	MIT
	SCHOLAR		Bolt, Beranek and Newman

TABLE 6-3. Existing Expert Systems by Function (cont.)

Function	Domain	System[a]	Institution
Knowledge acquisition	Medical diagnosis	TEIRESIAS	Stanford Univ.
	Medical consultation	EXPERT	Rutgers Univ.
	Geology	KAS	SRI
Expert system construction		ROSIE	Rand
		AGE	Stanford Univ.
		Hearsay III	Univ. of Southern California/ ISI
		EMYCIN	Stanford Univ.
		OPS 5	Carnegie-Mellon Univ.
		YES/MVS	IBM
	Medical diagnosis	KMS	Univ. of Maryland
	Medical consultation	EXPERT	Rutgers Univ.
	Electronic systems diagnosis	ARBY	Smart Systems Technology
	Medical consultation using time-oriented data	MECS-AI	Tokyo Univ.
Consultation/intelligent assistant	Battlefield weapons assignments	BATTLE	NRL AI Lab.
	Medicine	Digitalis Therapy Advisor	MIT
	Radiology	RAYDEX	Rutgers Univ.
	Computer sales	XCEL	Carnegie-Mellon Univ./Digital Equipment Corp.
	Medical treatment	ONCOCIN	Stanford Univ.
	Nuclear power plants	CSA Model-Based Nuclear Power Plant Consultant	Georgia Tech.
	Diagnostic prompting in medicine	RECONSIDER	Univ. of California, San Francisco
Management	Automated factory	IMS	Carnegie-Mellon Univ.
	Project management	CALLISTO	Digital Equipment Corp.

Automatic programming	Modeling of oil well logs	ΦNIX	Schlumberger-Doll Research
		CHI	Kestrel Inst.
		PECOS	Stanford Univ.
		LIBRA	Stanford Univ.
		SAFE	University of Southern California/ISI
		DEDALUS	SRI
		Programmer's Apprentice	MIT
Image understanding		VISIONS	Univ. of Massachussetts
		ACRONYM	Stanford Univ.

[a] References to these systems can be found in Duda (1981), Stefik et al. (1982), Buchanan (1981), Buchanan and Duda (1982), Barr and Feigenbaum (1982), *IJCAI-81*, *IJCAI-83*, *AAAI-82*, *AAAI-83*, and *AAAI-84*.

- *Building and Construction*
    Design, planning, scheduling, control
- *Equipment*
    Design, monitoring, control, diagnosis, maintenance, repair, instruction.
- *Command and Control*
    Intelligence analysis, planning, targeting, communication
- *Weapon Systems*
    Target identification, adaptive control, electronic warfare
- *Professions*
    (Medicine, law, accounting, management, real estate, financial, engineering)
        Consulting, instruction, analysis
- *Education*
    Instruction, testing, diagnosis, concept formation and new knowledge development from
        experience.
- *Imagery*
    Photo interpretation, mapping, geographic problem-solving.
- *Software*
    Instruction, specification, design, production, verification, maintenance
- *Home Entertainment and Advice-giving*
    Intelligent games, investment and finances, purchasing, shopping, intelligent information
        retrieval
- *Intelligent Agents*
    To assist in the use of computer-based systems
- *Office Automation*
    Intelligent systems
- *Process Control*
    Factory and plant automation
- *Exploration*
    Space, prospecting, etc.

Figure 6-4   Future opportunities for expert systems.

On a more near-term scale, in the next few years we will increasingly see expert systems with thousands of rules. In addition to the increasing number of rule-based systems we can also expect to see an increasing number of non-rule-based systems. Also anticipated are much improved explanation systems that can explain (make "transparent") why an expert system did what it did and what things are of importance. By the late 1980s we can expect to see intelligent, friendly, and robust human interfaces and much better system building tools, a big improvement over the commercial expert system building tools proliferating in the mid 1980s. Somewhere around the year 2000 we can expect to see the beginnings of systems that semi-autonomously develop knowledge bases from text. The result of these developments may very well herald a maturing information society where expert systems provide the capability for putting experts at everyone's disposal. In the process, production and information costs should greatly diminish, opening up major new opportunities for societal betterment.

REFERENCES

AAAI-82—Proceedings of the National Conference on Artificial Intelligence, Carnegie-Mellon Univ. and Univ. of Pittsburgh, Pittsburgh, PA, Aug. 18–22, 1982.

AAAI-83—Proceedings of the National Conference on Artificial Intelligence, Washington, DC, Aug. 22–26, 1983.

AAAI-84—Proceedings of the National Conference on Artificial Intelligence, Univ. of Texas, Austin, TX, Aug. 6–10, 1984.

Barr, A., and Feigenbaum, E. A., *The Handbook of Artificial Intelligence,* Vol. 2. Los Altos, CA: W. Kaufmann, 1982.

Boose, J. H., "Personal Construct Theory and the Transfer of Human Expertise," *AAAI-84,* Aug. 1984, pp. 27-33.

Buchanan, B. G., "Research on Expert Systems," Report STAN-CS-81-837, Dept. of Computer Science, Stanford Univ., 1981.

Buchanan, B. G., and Duda, R. O., "Principles of Rule-Based Expert Systems," Heuristic Programming Project Report HPP 82-14, Dept. of Computer Science, Stanford Univ., Stanford, CA, Aug. 1982. (To appear in *Advances in Computers,* Vol. 22, M. Yorit (Ed.). New York: Academic Press.)

Chandrasekaran, B., "Towards a Taxonomy of Problem Solving Types," *AI Magazine,* Vol. 4, No. 1, Winter–Spring 1983, pp. 9–17.

Davis, R., "Expert Systems: Where Are We: and Where Do We Go from Here?" *AI Magazine,* Vol. 3, No. 2, Spring 1982, pp. 3–25.

Duda, R. O., "Knowledge-Based Expert Systems Come of Age," *Byte,* Vol. 6, No. 9, Sept. 1981, pp. 238-281.

Feigenbaum, E. A., "Knowledge Engineering for the 1980s," Dept. of Computer Science, Stanford Univ., Stanford, CA, 1982.

Feigenbaum, E. A., Buchanan, B. G., and Lederberg, J., "On Generality and Problem Solving: A Case Study Using the DENDRAL Program," in *Machine Intelligence 6,* B. Meltzer, and D. Michie (Eds.). New York: Wiley, 1971, pp. 165–190.

Gevarter, W. B., *An Overview of Expert Systems,* NBSIR 82-2505, National Bureau of Standards, Washington, DC, May 1982 (rev. Oct. 1982).

Gevarter, W. B., "Expert Systems: Limited but Powerful," *Spectrum,* Aug. 1983, pp. 39–45.

IJCAI-81—Proceedings of The International Joint Conference on Artificial Intelligence, Vancouver, Aug. 1981.

IJCAI-83—Proceedings of The Eighth International Joint Conference on Artificial Intelligence, Karlsruhe, West Germany, Aug. 1983.

Lenat, D. B., "AM: An Artificial Intelligence Approach to Discovery in Mathematics as Heuristic Search," Ph.D. dissertation, Memo AIM-286, AI Lab., Stanford Univ., Stanford, CA, 1976.

Lindsay, R. K., Buchanan, B. G., Feigenbaum, E. A., and Lederberg, J., *Applications of Artificial Intelligence for Organic Chemistry: The DENDRAL Project,* New York: McGraw-Hill, 1980.

McDermott, J., "R1: A Rule-Based Configurer of Computer Systems," *Artificial Intelligence*, Vol. 19, 1982, pp. 39-88.

Miller, R. A., Pople, H. E., and Myers, J. D., "Internist-I, An Experimental Computer-Based Diagnostic Consultant for General Internal Medicine," *The New England Journal of Medicine*, Vol. 307, No. 8, Aug. 19, 1982, pp. 468-476.

Shortliffe, E. H., *Computer-Based Medical Consultations: MYCIN*. New York: American Elsevier, 1976.

Stefik, M., Alkins, J., Balzer, R., Benoit, J., Birnbaum, L., Hayes-Roth, R., and Sacerdoti, E., "The Organization of Expert Systems, A Tutorial," *Artificial Intelligence*, Vol. 18, 1982, pp. 135-173.

Stefik, M., Bobrow, D. G., Mittal, S., and Conway, L., "Knowledge Programming in LOOPS," *AI Magazine*, Vol. 4, No. 3, Fall 1983, pp. 3-13.

7

PLANNING

7-1. INTRODUCTION

Nilsson at SRI originally specified problem solving and planning as being one of the four fundamental application areas of AI. However, the *weak methods*, employing little domain knowledge, originally used in AI for problem solving and planning, proved inadequate for complex real-world problems. Thus in seeking solutions in this area, larger amounts of knowledge have since been utilized. The net result has been that the knowledge engineering methodology used for expert systems has been adapted for use in problem solving and planning. Thus the boundary between problem solving and planning and expert systems has faded and it is now common to refer to all these knowledge-based activities as expert systems, and they were therefore covered in Chapter 6. Nevertheless, this chapter reviews briefly some of the earlier, less knowledge-intensive planning systems and several examples of recent systems.

Most AI applications can be considered as examples of problem solving, which are well covered in the other AI application areas: expert systems, computer vision, language understanding, and so on. In this chapter we consider only planning systems. Planning can be defined for our purposes as the design process for selecting and stringing together individual actions into sequences in order to achieve desired goals.

7-2. BASIC PLANNING PARADIGM

Wilensky (1983) outlines the basic structure of plans from the viewpoints of commonsense problem solving and natural language understanding. A schematic for
Wilensky's basic planning paradigm is given in Fig. 7-1. In this paradigm the planner
recognizes from the environment that a new situation has arisen which merits a

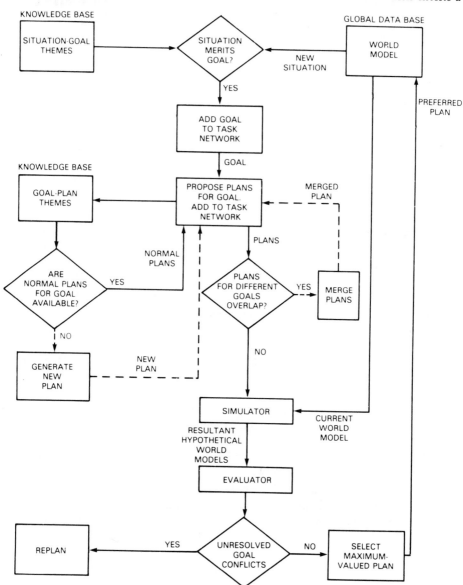

Figure 7-1 Wilensky planning paradigm.

goal. The planner then retrieves from memory a plan that might be used to achieve this goal, or generates a new trial plan if no existing plan is suitable. This candidate plan is then projected forward (via simulation) to observe the outcome. This outcome is examined to see if there are any conflicts that will arise in achieving other goals if this plan is pursued. If not, this and other candidate plan outcomes are evaluated and the maximum-valued plan is chosen. The plan, when implemented, will modify the current state of affairs. This impact, together with any other changes in the environment, results in a new world model with new situations that may merit new goals, so that the cyclic process of planning continues. When candidate plans are being considered, if the candidate plan overlaps existing plans for other goals, these overlapping plans may be merged to conserve resources.

A basic problem in planning is that of conflicting goals. The causes of conflicting goals are indicated in Fig. 7-2. (A preservation goal is a goal to preserve an already existing condition, or is a goal not to undo a desirable state or goal resulting from another plan.)

Problems arising from conflicting goals are dealt with by replanning or by eliminating the factors causing the goal conflicts. A flow diagram for resolving goal conflicts is given in Fig. 7-3. If the goal conflicts cannot be completely resolved, partial fulfillment of goals may be attempted or goals of lesser importance may have to be dropped. The global strategy is to achieve as many goals as possible, maximizing the composite value of the goals achieved, and not waste resources in achieving them.

DEVISER (Vere, 1983) is a good example of a planning program designed to deal with conflicting goals resulting from resource and time constraints. Wilensky also discusses "competing goals" that arise in competitive situations. The planning strategies given in this case are to:

1. Avoid conflicts
2. Outdo an opponent
3. Hinder an opponent
4. Induce alterations in competitor's plans

7-3. PARADIGMS FOR GENERATING PLANS

The major issue in any planning system is reducing search. The other key issue is how to handle interacting subproblems. The following paradigms are different approaches to addressing these issues.

Cohen and Feigenbaum (1982) discuss four distinct approaches to planning: nonhierarchical, hierarchical, script-based (skeletal), and opportunistic. (Figure 7-4 indicates key planning systems that have evolved over the years based on these approaches.) Virtually all plans, both hierarchical and nonhierarchical, have hierarchical subgoal structures. That is, each goal can be expanded into several subgoals, which themselves can be further expanded, and so on, until the bottom level consists of operators needed to achieve the lowest-level goals. The distinction between hierarchical and nonhierarchical planners is that "a hierarchical planner generates a hierarchy of representations of a plan in which the highest is a simplification, or abstraction of the plan and the lowest is a detailed plan, sufficient to solve the problem. In contrast, nonhierarchical planners have only one representation of a plan" (pp. 516–517).

7-3.1. Nonhierarchical Planning

Nonhierarchical planning does not initially distinguish between important and unimportant actions, so that everything is considered in the initial plan, including cumbersome details. For complex problems, this often results in a large search. One way the search can be greatly reduced is by initially assuming subgoals independent

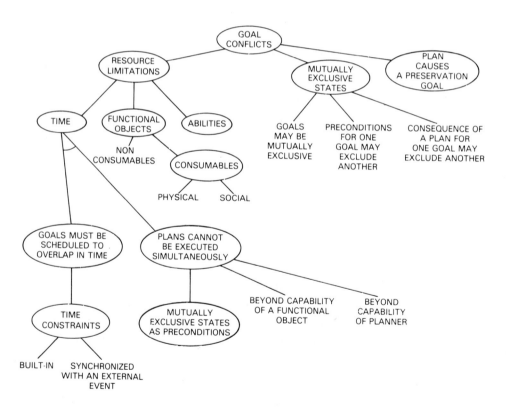

Figure 7-2 Nature of goal conflicts.

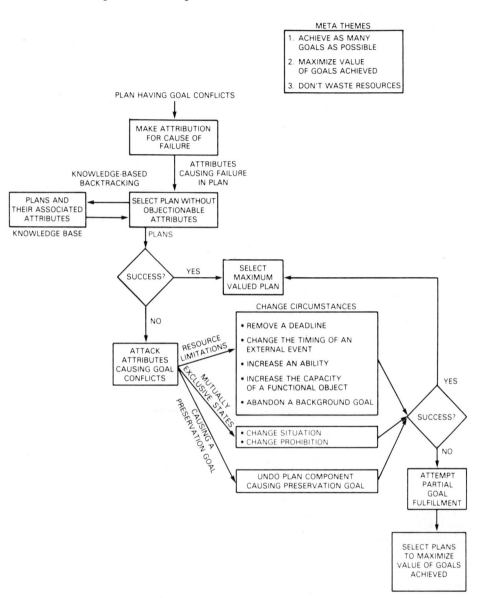

Figure 7-3 Resolving conflicting goals by replanning (and/or attacking factors causing conflicts).

and then trying to repair the plan to account for the interactions (as in HACKER, Table 7-1A*).

A knowledge-based approach used in ISIS-II (Fox et al., 1982, Table 7-1B) is

*Tables 7-1A through 7-1F use the expert systems format developed in Chapter 6.

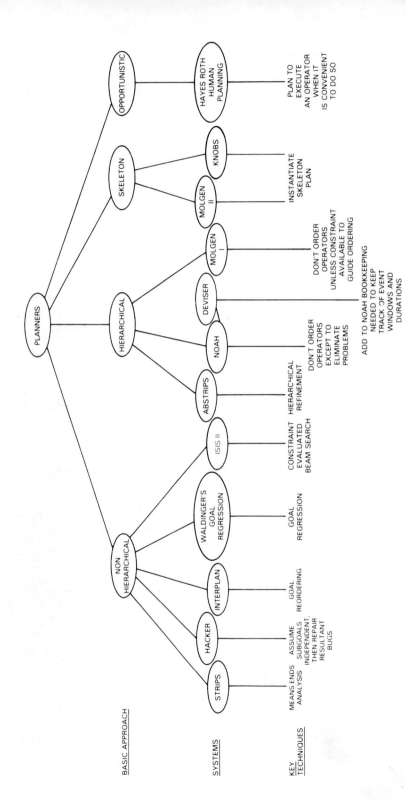

Figure 7-4 Planning techniques.

TABLE 7-1A Planners: HACKER

System: HACKER
Institution: MIT
Author: Sussman (1975)

| | | | Key Elements of: | |
Purpose	Approach	Knowledge Base	Global Data Base	Control Structure
Skill acquisition: devises a skill (set of procedures) to solve a problem, for example, plan to reorder blocks on a stack	Formulate plans to solve subgoals independently and then patch them up (e.g., to correct interferences where achieving one subgoal may prevent accomplishment of another) Solves problems by: 1. Searching for an appropriate procedure 2. If procedure does not achieve desired goal, reasons for failure are formalized as bugs 3. Using library of bug correction procedures, the plan is debugged If no procedure is available to solve a problem, a new procedure is written using the programming techniques library	Answer library: problem-solving procedures Knowledge library: facts about the domain Programming techniques library: to devise new problem-solving procedures Library of generic bugs Library of bug correction procedures	Goals Procedures used Bugs	Search for appropriate procedures to achieve goals and correct bugs Write new procedure when no appropriate procedure is found

73

TABLE 7-1B Planners: ISIS-II

System: ISIS-II
Institution: Carnegie-Mellon University
Authors: Fox, et al. (1982)

Purpose	Approach	Knowledge Base	Global Data Base	Control Structure
			Key Elements of:	
Job-shop planning/ scheduling of parts production	Generate schedules by heuristic search using evaluation functions based on constraints associated with costs, process applicability, machine availability, and supervisor preferences Set up mechanism to dynamically relax constraints as required **Sequence for generating a schedule** 1. Use constraints to perform a rule-based presearch analysis to bound search 2. Do a constraint-directed beam-search where only the top-rated n partial paths are saved 3. Perform postsearch analysis to determine if search was effective	Constraints (and their importance) Organization goals (associated with profit) Physical constraints Gating constraints (preconditions for object applicability or process initiation)	Preference constraints Queue positions Machine preferences Partial paths and their evaluations Work in progress Shop status Goals, due dates, and attributes of parts to be manufactured	Presearch pruning of search space—based on constraints Beam search using evaluation functions based on constraints

to prune the search space prior to search by using constraints, and then narrow the space actually searched by using a *beam search* approach.

7-3.2. Hierarchical Planning

In this approach, first a high-level plan is formulated considering only the important aspects; then the vague parts of the plan are refined into more detailed subplans. By ignoring the details at the higher levels, search is vastly reduced. ABSTRIPS (Table 7-1C) is illustrative of this approach where only the top-rated partial paths are considered.

7-3.3. Utilization of Skeleton Plans

This approach utilizes stored plans which contain the outlines for solving many different kinds of problems. The skeleton plans are then filled in for the particular problem being solved. This technique has similarities to Schank's script-based approach to language understanding discussed in Chapter 9. KNOBS (Engelman et al., 1980, Table 7-1D), a frame-based planning system for tactical air strikes, is an example of a skeletal plan approach.

7-3.4. Opportunistic Planning

Opportunistic planning (Hayes-Roth and Hayes-Roth, 1978) is based on the way that human beings often approach planning. In this approach, the plan is developed piecewise, with parts of the plan being developed separately, and then added to, enlarged, and linked together as opportunities present themselves. Planning of this sort incorporates both top-down and bottom-up components.

Planners evolve by building on past techniques. For example, DEVISER (Table 7-1F), the first planner to deal explicitly with time, is based on NOAH (Table 7-1E), with facilities having been added to keep track of event "windows" and durations. Figure 7-5 presents a simplified flowchart of DEVISER's core planning component.

7-4. CURRENT STATE OF THE ART

Although over a dozen research planning systems have been developed, AI planners are just now (1984) beginning to enter actual use. Examples of planners approaching operational use are DEVISER, ISIS-II, and KNOBS.

7-5. PLAYERS AND RESEARCH TRENDS

Principal developers of planning systems have been SRI, Stanford University, MIT, Rand, Carnegie-Mellon University, MITRE, and JPL. Current research is focused on

TABLE 7-1C Planners: ABSTRIPS

System: ABSTRIPS
Institution: SRI
Author: Sacerdoti (1974)

Purpose	Approach	Key Elements of:		Control Structure
		Knowledge Base	Global Data Base	
Devises plans for a robot to move objects between rooms	Do hierarchical planning by first devising a top-level plan based on the key aspects of the problem, then successively refining it by considering less critical aspects of the problem 1. Fix abstraction levels for solutions (plans) 2. Problem solution proceeds top down (most abstract to most specific) 3. Complete solution at one level and then move to next level below	Criticality assignments of elements in robot planning domain Configuration of the rooms Objects and their properties in the domain Rules for decrementing criticality level Heuristic search rules for each level	Goal Initial state of system (criticality at maximum) Plans thus far Current criticality level	Goal directed (backward chaining at each level) Top-down refinement of plans using hierarchial abstract search spaces

System: KNOBS
Institution: MITRE
Author: Engelman (1980)

TABLE 7-1D Planners: KNOBS

Purpose	Approach	Key Elements of:		
		Knowledge Base	Global Data Base	Control Structure
Planning consultant for Air Force tactical missions Other domains include: Naval "show of flag" missions Scheduling of crew activities for the NASA space shuttle	Assist a user by interactively accepting mission data and using it to instantiate a stereotypical solution to user's problem—checking input for inconsistencies and oversights Represent the stereotypical missions as frames; the checks are constraints among the possible slot values in such frames Uses a natural language interface (APE II)	Targets stored hierarchically in frames—individual targets inherit from generic targets Frames representing prototypical missions and submissions Resource data—frames representing static descriptions of object attributes, with inheritances via linkage to more generic frames Scripts composed of causally linked chains Overall knowledge base network consists of several thousand frames Rules for instantiation of frames and slots	Target Airbase from which to fly mission Type of aircraft Armaments **Plan thus far** etc.	Frame instantiation uses rules and constraints Backward chaining of production rules in a MYCIN-like deductive manner to manage such generic choices as aircraft, weapons, support, and electronic countermeasures Inference mechanism uses a syntactic pattern matcher with provisions for restrictions on variable instantiations

77

TABLE 7-1E Planners: NOAH

System: NOAH
Institution: SRI
Authors: Sacerdoti (1975)

Purpose	Approach	Key Elements of:		
		Knowledge Base	Global Data Base:	Control Structure
Robot planning system (assigns an ordering to operators in a plan, e.g., an assembly task)	Hierarchical planner—develops hierarchy of subgoals by expanding goals (lowest-level subgoals eventually expanded by problem-solving operators) Expands, in parallel, individual plans for interacting subgoals, but initially assigns only a partial ordering to operators; stops when interference between the partial subgoal plans is observed, and adjusts the ordering of the operators as needed to resolve the interference Develops procedural nets to represent plans as they are developed	Rules for recognizing interference between plans Rules for resolving interferences Domain knowledge Functions that expand goals into subgoals Operators to transform one state to another; effects of actions are represented explicitly (via add lists and delete lists)	World model Goal Subgoals Partial ordering of operators in subgoal plans Interference between plans	Least commitment Backward chaining

dealing with time, interacting subgoals, planning techniques, and dealing with constraints.

7-6. FUTURE DIRECTIONS

Automatic planning is still a difficult task. The current trend is toward the use of knowledge engineering to configure planners as expert systems. Thus knowledge-based planners were included in Chapter 6. Another trend is toward increased concern with spatial-temporal planning. This is exemplified by Malik and Binford (1983), Allen and Koomen (1983), and Brooks (1983).

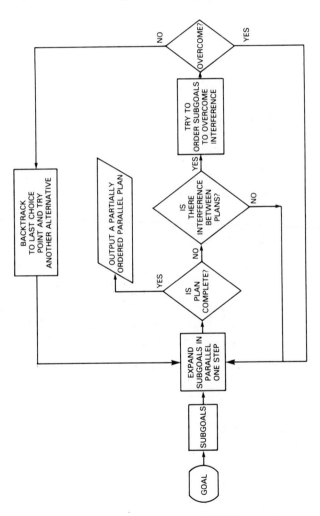

Figure 7-5 Simplified flowchart of DEVISER's core planner.

TABLE 7-1F Planners: DEVISER

System: DEVISER
Institution: JPL
Author: Vere (1983)

Purpose	Approach	Knowledge Base	Key Elements of: Global Data Base	Control Structure
General-purpose automated planner/scheduler to generate parallel plans to achieve goals with time constraints (e.g., scheduling spacecraft actions during a planetary flyby)	Backward chaining from unordered subgoals by: 1. Satisfying goals, where possible, by linking goal nodes with the same already achieved nodes 2. If subgoals cannot be met by linking, nodes are expanded in parallel, step by step, into activities which achieve the subgoals 3. When two parallel expansions produce contradictions, conflicts are resolved by ordering nodes (formerly unordered) 4. If conflicts cannot be resolved by ordering, DEVISER backtracks to the last choice point and tries another alternative A window start time for each activity in the plan is updated dynamically during plan generation, in order to maintain consistency with the windows and durations of adjacent goals and activities	Rules for recognizing interferences between subgoal expansions Rules for reordering subgoal plans to resolve conflicts Domain knowledge: Operators to transform one state to another; effects of actions are represented explicitly by add lists and delete lists Goal windows and durations Event schedules	World model Subgoals Ordering of operators in subgoal plans Interferences between subgoal plans Node expansion histories Current windows	Least commitment Backward chaining Dynamic maintenance of windows of activities and goals to preserve consistency

REFERENCES

Allen, J. F., and Koomen, J. A., "Planning Using a Temporal World Model," in *Proceedings of the Eighth International Joint Conference on Artificial Intelligence*, Karlsruhe, West Germany, Aug. 8–12, 1983. Los Altos, CA: W. Kaufmann, 1983, pp. 741–747.

Brooks, R. A., "Find-Path for a PUMA-Class Robot," in *Proceedings of the National Conference on Artificial Intelligence*, Washington, DC, Aug. 22–26, 1983. Los Altos, CA: W. Kaufmann, 1983, pp. 40–44.

Cohen, P. R., and Feigenbaum, E. A., *The Handbook of Artificial Intelligence*, Vol. 3. Los Altos, CA: W. Kaufmann, 1982.

Engelman, C., Scarl, E., and Berg, C., "Interactive Frame Instantiation," in *Proceedings of the First Annual Conference on Artificial Intelligence*, Stanford, CA, Aug. 1980.

Fox, M. S., Allen, B., and Strohm, G., "Job-Shop Scheduling: An Investigation in Constraint-Directed Reasoning," in *Proceedings of the Second Annual National Conference on Artificial Intelligence,* Carnegie-Mellon Univ., Aug. 1982, pp. 155–158.

Hayes-Roth, B., and Hayes-Roth, F., "Cognitive Processes in Planning," Rep. R-2366-ONR, Rand Corp., Santa Monica, CA, 1978.

Malik, J., and Binford, T O., "Reasoning in Time and Space," in *Proceedings of the Eighth International Joint Conference on Artificial Intelligence*, Karlsruhe, West Germany, Aug. 8–12, 1983. Los Altos, CA: W. Kaufmann, 1983, pp. 343–345.

Sacerdoti, E. D., "Planning in a Hierarchy of Abstraction Spaces," *Artificial Intelligence*, Vol. 5, 1974, pp. 115–135.

Sacerdoti, E. D., *A Structure for Plans and Behavior*, Tech. Note 109, AI Center, SRI International, Menlo Park, CA, 1975.

Sussman, G. J., *A Computer Model of Skill Acquisition.* New York: Elsevier, 1975.

Vere, S., "Planning in Time: Windows and Durations for Activities and Goals," *IEEE Transactions on Pattern Analysis and Machine Intelligence*, Vol. PAMI-5, No. 3, May 1983, pp. 246–266.

Wilenksy, R., *Planning and Understanding: A Computational Approach to Human Reasoning.* Reading, MA: Addison-Wesley, 1983.

8

COMPUTER VISION

8-1. INTRODUCTION

Computer vision—visual perception employing computers—shares with expert systems the role of being one of the most popular topics in artificial intelligence today. The computer vision field is multifaceted, having many participants with diverse viewpoints, with many papers having been written. However, the field is still in the early stages of development—organizing principles have not yet fully crystallized and the associated technology has not yet been completely rationalized. However, commercial vision systems have already begun to be used in manufacturing and robotic systems for inspection and guidance tasks, and other systems (at various stages of development) are beginning to be employed in military, cartographic, and image interpretation applications. Examples of current applications are indicated in Table 8-1.

Computer (computational or machine) vision can be defined as perception by a computer based on visual sensory input. Barrow and Tenenbaum (1981, p. 573) state:

> Vision is an information-processing task with well-defined input and output. The input consists of arrays of brightness values, representing projections of a three-dimensional scene recorded by a camera or comparable imaging device. Several input arrays may provide information in several spectral bands (color) or from multiple viewpoints (stereo or time sequence). The desired output is a concise description of the three-dimensional scene depicted in the image,

TABLE 8-1   Examples of Applications of Computer Vision Now Under Way

---

Automation of industrial processes
   Object acquisition by robot arms, for example, for sorting or packing items arriving on
     conveyor belts
   Automatic guidance of seam welders and cutting tools
   VLSI (very large scale integration)-related processes, such as lead bonding, chip alignment,
     and packaging
   Monitoring, filtering, and thereby containing the flood of data from oil drill sites or from
     seismographs
   Providing visual feedback for automatic assembly and repair

Inspection tasks
   Inspection of printed circuit boards for spurs, shorts, and bad connections
   Checking the results of casting processes for impurities and fractures
   Screening medical images such as chromosome slides, cancer smears, x-ray and ultrasound
     images, tomography
   Routine screening of plant samples
   Inspection of alphanumerics on labels and manufactured items
   Checking packaging and contents in pharmaceutical and food industries
   Inspection of glass items for cracks, bubbles, etc.

Remote sensing
   Cartography, automatic generation of hill-shaded maps, and registration of satellite images
     with terrain maps
   Monitoring traffic along roads, docks, and at airfields
   Management of land resources such as water, forestry, soil erosion, and crop growth
   Detecting mineral ore deposits

Making computer power more accessible
   Management information systems that have a communication channel considerably wider
     than current systems that are addressed by typing or pointing
   Document readers (for those who still use paper)
   Design aids for architects and mechanical engineers

Military applications
   Tracking moving objects
   Automatic navigation based on passive sensing
   Target acquisition and range finding

Aids for the partially sighted
   Systems that read a document and speak what they read
   Automatic "guide dog" navigation systems

---

the exact nature of which depends upon the goals and expectations of the
observer. It generally involves a description of objects and their interrelationships, but may also include such information as the three-dimensional
structures of surfaces, their physical characteristics (shape, texture, color,
material), and the locations of shadows and light sources.*

*Reprinted with permission from H. G. Barrow and J. M. Tenenbaum, "Computational
Vision," *Proceedings of the IEEE*, Vol. 69, No. 5, May 1981. Copyright © 1981 by the Institute of Electrical and Electronics Engineers, Inc.

8-2. RELATION TO HUMAN VISION

MIT's Marr and Nishihara (1978, p. 42) take the view that "artificial intelligence is (or ought to be) the study of information processing problems that characteristically have their roots in some aspects of biological information processing." They developed a computational theory of vision based on their study of human vision. Figure 8-1 represents the transition from the raw image through the primal sketch to the 2½-dimensional sketch (exemplified by Fig. 8-2), which contains information on local surface orientations, boundaries, and depths.

The primal sketch, reminiscent of an artist's hurried drawing, is a primitive but rich description of the way the intensities change over the visual field. It can be represented by a set of short line segments separating regions of different brightness. [Figure 8-3(b) can be considered to be an example of a primal sketch.] A list of the properties of line segments, such as location, length, and orientation for each segment, can be used to represent the primal sketch. The late D. C. Marr and his associates' development of a human visual information-processing theory (Marr, 1982) has had a substantial impact on computational vision.

There are strong indications (see, e.g., Gevarter, 1977) that the interpretative planning areas of the human brain set up a context for processing the input data. [This viewpoint is captured by Minksy's (1975) AI "frame" concept for knowledge representation.] The brain then uses visual and other cues from the environment to draw in past knowledge to generate an internal representation and interpretation of the scene. This knowledge-based expectation-guided approach to vision is now appearing in advanced AI computer vision systems.

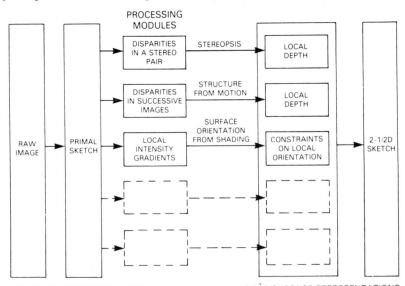

Figure 8-1 Framework for early and intermediate states in a theory of visual information processing. (From Marr and Nishihara, 1978, p. 42. Reprinted with permission from *Technology Review*, copyright 1978.)

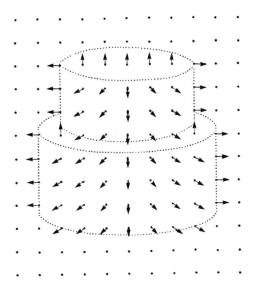

A candidate for the so-called 2-½-dimensional sketch, which encompasses local determinations of the depth and orientation of surfaces in an image, as derived from processes that operate upon the primal sketch or some other representation of changes in gray-level intensity. The lengths of the needles represent the degree of tilt at various points in the surface; the orientations of the needles represent the directions of tilt... Dotted lines show contours of surface discontinuity. No explicit representation of depth appears in this figure.

Figure 8-2   Example of a 2½ D sketch. (From Marr and Nishihara, 1978, p. 41. Reprinted with permission from *Technology Review*, copyright 1978.)

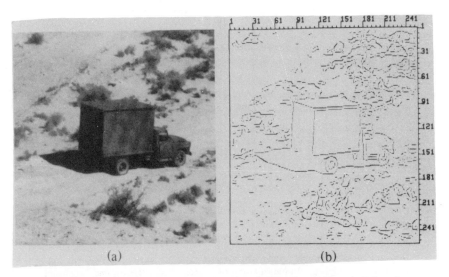

Figure 8-3   (a) Truck image and (b) edges detected in it by the Nevatia-Babu edge detector. (From Ramakant Nevatia, *Machine Perception*, © 1982, p. 114. Reprinted by permission of Prentice-Hall, Inc., Englewood Cliffs, N.J.)

8-3. APPROACHES TO COMPUTER VISION

8-3.1. Basis for a General-Purpose Image Understanding System

Barrow and Tenenbaum (1981, p. 573) observe that in going from a scene to an image (an array of brightness values) the image encodes much information about the scene, but the information is confounded in the single brightness value at each point. In projecting onto the two-dimensional image, information about the three-dimensional structure of the scene is lost. To decode brightness values and recover a scene description, it is necessary to employ a priori knowledge embodied in models of the scene domain, the illumination, and the imaging process.

As indicated by Fig. 8-4, computer vision is an active process that uses these models to interpret the sensory data. To accommodate the diversity of appearance found in real imagery, a high-performance, general-purpose system must embody a great deal of knowledge in its models.

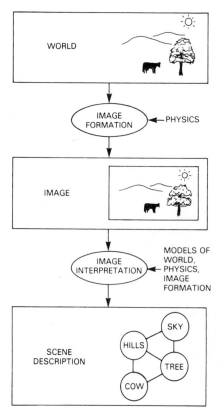

Figure 8-4 Model-based interpretation of images. (From Barrow and Tenebaum, 1981, p. 573. Reprinted with permission. Copyright © 1981 by the Institute of Electrical and Electronic Engineers, Inc.)

8-3.2. Basic Paradigms for Computer Vision*

In broad terms, an image-understanding system starts with the array of picture-element amplitudes that define the computer image, and using stored models (either specific or generic) determines the content of a scene. Typically, various symbolic features such as lines and areas are first determined from the image. These are then compared with similar features associated with stored models to find a match when specific objects are being sought. In more generic cases, it is necessary to determine various characteristics of the scene, and using generic models, determine from geometric shapes and other factors (such as allowable relationships between objects) the nature of the scene content.

A variety of paradigms have been proposed to accomplish these tasks in image-understanding systems. These paradigms are based on a common set of broadly defined processing and manipulating elements: feature extraction, symbolic representation, and semantic interpretation. The paradigms differ primarily in how these elements (defined below) are organized and controlled, and the degree of artificial intelligence and knowledge employed.

Hierarchical bottom-up approach.   Figure 8-5(A) is a block diagram of a hierarchical paradigm of an image-understanding system that employs a bottom-up processing approach. First, primitive features are extracted from the array of picture-element intensities that constitute the observed image. Examples of such features are picture-element ("pixel") amplitudes, edge point locations, and textural descriptors.

Next, this set of features is passed on to the semantic interpretation stage, where the features are grouped into symbolic representations. For example, edge points are grouped into line segments or closed curves, and adjacent region segments of common attributes are combined. The resultant symbol set of lines, regions, and so on, in combination with a priori stored models, are then operated on (i.e., semantically interpreted) to produce an application-dependent scene description.

Bottom-up refers to the sequential processing and control operation of the system starting with the input image. The key to success in this approach lies in a sequential reduction in dimensionality from stage to stage. This is vital because the relative processing complexity is generally greater at each succeeding stage. The hierarchical bottom-up approach can be developed successfully for domains with simple scenes made up of only a limited number of previously known objects.

Hierarchical top-down approach.   This approach (usually called *hypothesize and test*), shown in Fig. 8-5(B), is goal directed, the interpretation stage being guided in its analysis by trial or test descriptions of a scene. An example would be using template matching—*matched filtering*—to search for a specific object or structure within the scene. Matched filtering is normally performed at the pixel level by cross-

*This section is based primarily on Pratt (1978, pp. 570–574).

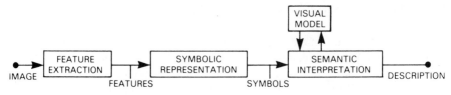

A. HIERARCHICAL BOTTOM-UP APPROACH

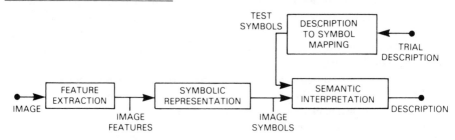

B. HIERARCHICAL TOP-DOWN APPROACH

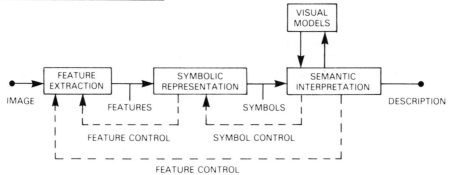

C. HETERARCHICAL APPROACH

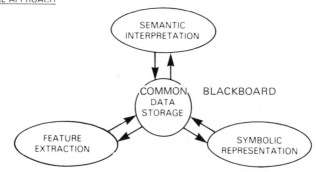

D BLACKBOARD APPROACH

Figure 8-5 Basic image-understanding paradigms. (From Pratt, 1978, pp. 570–574. Reprinted with permission. Copyright © 1978 by John Wiley & Sons, Inc.)

correlation of an object template with an observed image field. It is often computationally advantageous, because of the reduced dimensionality, to perform the interpretation at a higher level in the chain by correlating image features or symbols rather than pixels.

Heterarchical approach.   Hierarchical image-understanding systems are normally designed for specific applications. They thus tend to lack adaptability. A large amount of processing is also usually required. Pratt (1978, pp. 572-573) observes that often much of this processing is wasted in the generation of features and symbols not required for the analysis of a particular scene. A technique to avoid this problem is to establish a central monitor to observe the overall performance of the image-understanding system and then issue commands to the various system elements to modify their operation to maximize system performance and efficiency. Figure 8-5(C) is a block diagram of an image-understanding system that achieves heterarchical operation by distributed feedback control. An early heterarchical system for line finding in the limited domain of polyhedral objects was developed by Shairai (1975). Shairai's system detected the outer boundaries first (being of high contrast and easily detected) and then proposed hypotheses for the presence of other lines. These interior lines were then verified or rejected by a more sensitive edge detector.

Blackboard approach.   Another image-understanding system configuration, called the *blackboard model*, has been proposed by Reddy and Newell (1975). Figure 8-5(D) is a simplified representation of this approach in which the various system elements communicate with each other via a common working data storage called the *blackboard*. Whenever any element performs a task, its output is put into the common data storage, which is independently accessible by all other elements. The individual elements can be designed to act autonomously to further the common system goal as required. The blackboard system is particularly attractive in cases where several hypotheses must be considered simultaneously and their components need to be kept track of at various levels of representation.

8-3.3. Levels of Representation

A computer vision system, like human vision, is commonly considered to be naturally structured as a succession of levels of representation. Tenenbaum et al. (1979, pp. 254-255) sketch a way in which to view an organization of a general-purpose vision system (Fig. 8-6). They divide the figure into two parts. The first is image oriented (iconic), domain independent, and based on the image data (data driven). The second part of the figure is symbolic, dependent on the domain and the particular goal of the vision process.

The first portion takes the image, which consists of an intensity array of picture elements (pixels, e.g., 1000 X 1000), and converts it into image features such as edges and regions. These are then converted into a set of parallel *intrinsic images*, one each for distance (range), surface orientation, reflectance,* and so on.

*Fraction of normal incident illumination reflected.

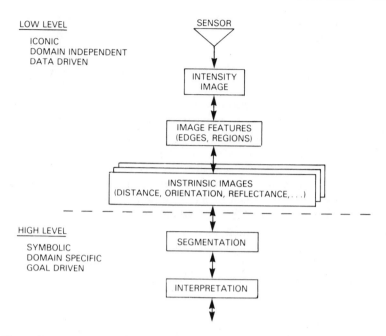

Figure 8-6 Organization of a visual system. (From Tenenbaum et al., 1979, p. 255. Reprinted with permission of Plenum Press, New York.)

The second part of the system segments these into volumes and surfaces dependent on the system's knowledge of the domain and the goal of the computation. Using domain knowledge and the constraints associated with the relations among objects in this domain, objects are identified and the scene analyzed consistent with the system goal.

8-4. CURRENT STATE OF THE ART

8-4.1. Human Vision

Human vision is the only available example of a general-purpose vision system. However, thus far, not many AI researchers have taken an interest in the computations performed by natural visual systems, but this situation is changing.

Many researchers believe that to a first approximation, the human visual system is subdivided into modules specializing in visual tasks. [For example, specific edge and orientation detectors have been isolated (Huebell and Wessel, 1979).] There is also evidence that people do global processing first and use it to constrain local processing.

Considerable information now exists about lower-level visual processing in human beings. However, as we progress up the human visual computing hierarchy, the exact nature of the appropriate representations becomes subject to dispute. Thus overall human visual perception is still very far from being understood.

8-4.2. Low and Intermediate Levels of Processing

Although methods for powerful high-level understanding visual analysis are still in the process of being determined, insights into low-level vision are emerging. The basic physics of imaging, and the nature of constraints in vision and their use in computation, are fairly well understood. Detailed programs for vision modules, such as *shape from shading** and shape from *optical flow* in an image (associated with object motion), have begun to appear. Also, the representational issues are now better understood.

However, even for well-understood low-level operations such as edge detection (see, e.g., Ballard and Brown 1982), there has been no convergence among the many techniques proposed, and no method stands out as the best. In general, edge detectors are still unreliable, although David Marr and Elaine Hildreth's approach at MIT, based on the zero crossing of the second derivative of the intensity gradient, is one technique that appears promising.

In industrial vision, the primary technique for achieving robust edge finding and segmentation is to use special lighting and convert to a silhouette binary image in which edges and regions are readily distinguishable. At intermediate levels, edge classification and labeling have been used very successfully in the blocks world.[†]

Binford (1982), in reviewing existing research in model-based vision systems, observed that most systems first segment regions, then describe their shape. None of the systems makes effective use of texture for segmentation and description. In general, shape description is primitive and interpretation systems have not yet made full use of even these limited capabilities.

As yet, the extraction of useful information from color is extremely rudimentary. The perceptual use of motion (optical flow) has been a focus of attention recently, but findings are preliminary.

For low-level processing, many recent algorithms take the form of parallel computations involving local interactions. One popular approach having this character is *relaxation*,[‡] in which local computations are iteratively propagated to try to extract global features. These locally parallel architectures are well suited to rapid parallel processing techniques using special-purpose VLSI chips.

---

[*]Barrow and Tenenbaum (1978) describe a low-level method of estimating relative distance and surface orientation from a single image. They use heuristics based on the rate of change of brightness across the image.

Ikeuchi and Horn (1981) have formulated a second-order differential equation which Horn calls the *image irradiance equation*. This equation relates the orientation of the local surface normal of a visible surface, its surface reflectance characteristics, and the lighting to the intensity value recorded at the corresponding point in the image.

[†]A small artificial world, consisting of blocks and pyramids, used to develop ideas in computer vision, robotics, and natural language interfaces.

[‡]The example starting on page 92 illustrates a relaxation process that could be implemented using parallel processing.

Example of Relaxation:* **Labeling of Edges to Determine Feasible Object Shapes From Line Images**

For trihedral polyhedra—objects for which exactly three plane surfaces come together at each vertex—the four possible vertex types are illustrated in Fig. B8-1. Assume that we label the image lines as follows:

1. A "+" line represents a convex edge with both of its planes visible from the camera.

2. A "−" line represents a concave edge with both of its planes visible from the camera.

3. A "←" or "→" line represents an occluding edge: a convex edge formed by one plane occluding the other as viewed from the camera. As one moves along the edge in the direction of the arrow, the occluding plane is to the right.

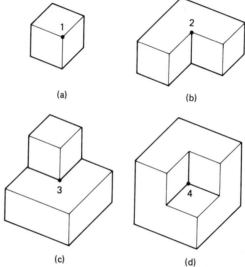

Figure B8-1 The four vertex types. (From Huffman, 1971.)

Then, for the vertex of the object in Fig. B8-1(a), we have three different labelings (Fig. B8-2) dependent on the direction from which the object is viewed.

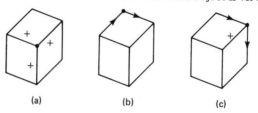

Figure B8-2 Three views of the vertex in Fig. B8-1(a).

Collecting the possible labelings of the four vertices in Fig. B8-1, we have Fig. B8-3.

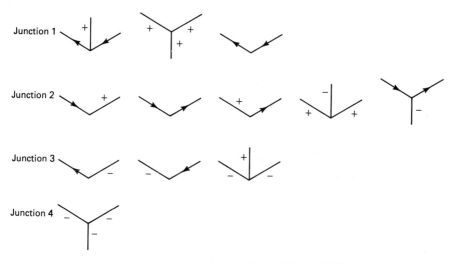

Figure B8-3   Junction labelings. (From Huffman, 1971.)

For the simple picture shown in Fig. B8-4, the possible labelings are as shown in Fig. B8-5.

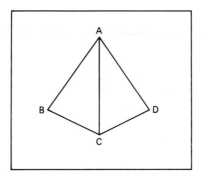

Figure B8-4   Simple picture. (From Clowes 1971.)

We will use the relaxation procedure called a *Waltz filtering algorithm* (Waltz, 1975). Starting with junction A in Fig. B8-4 and proceeding clockwise from A to D to C to B (choosing only those labels consistent with previous labels assigned by other joints to the lines coming into that joint), we have the results shown in Fig. B8-6. Note that each time labels are chosen for a joint (e.g., step 3) the inconsistent labels chosen for the previous joints are discarded (e.g., steps 4 and 5).

The resultant three possible labels of Fig. B8-4 are shown in Fig. B8-7. The interpretation of these labelings are: Fig. B8-7(c) is a pyramidlike object occluding a background; Fig. B8-7(a) and (b) are pyramidlike objects projecting from a sur-

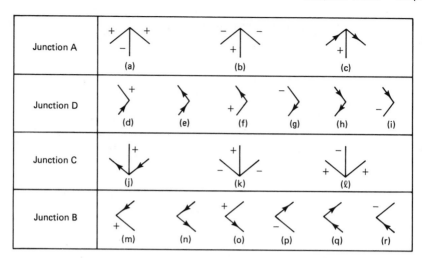

Figure B8-5 Possible labelings for the junctions in Fig. B8-4.

STEP	LABELS ASSIGNED TO JUNCTIONS			
	A	D	C	B
1				
2	Unchanged			
3	Unchanged	Unchanged		
4	Unchanged		Unchanged	
5		Unchanged	Unchanged	
6				

Figure B8-6 Label sets assigned to each junction during the steps of the constraint-satisfaction algorithm.

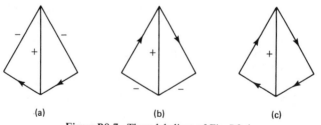

Figure B8-7 Three labelings of Fig. B8-4.

face, where the − edges are the concave edges corresponding to the junctures of the pyramids with the surfaces from which they are projecting.

8-4.3. Industrial Vision Systems

Barrow and Tenenbaum (1981, p. 572) observe that:

Significant progress has been made in recent years on practical applications of machine vision. Systems have been developed that achieve useful levels of performance on complex real imagery in tasks such as inspection of industrial parts, interpretation of aerial imagery, and analysis of chest x-rays. Virtually all such systems are special purpose, being heavily dependent on domain-specific constraints and techniques.*

It has been estimated that as of mid-1982, although fewer than 50 sophisticated industrial vision systems were actually in use in the United States, approximately 1000 simple line-scan inspection systems were in regular operation. Although special-purpose systems have thus far been the most effective, successful vision applications are now becoming commonplace and are expanding. Vision manufacturers are now beginning to provide easier user programming, friendlier user interfaces, and systems engineering support to prospective users. Many firms are now entering the industrial vision field, with technical leapfrogging being common due to rapidly changing technology.

8-4.4. General-Purpose Vision Systems

Although many practical image recognition systems have been developed, Hiatt (1981, pp. 2, 8) observes that "in current vision applications, the type of scene to be processed and acted upon is usually carefully defined and limited to the capability of the machine. . . . General purpose computer vision has not yet been solved in practice." This domain specificity makes each new application expensive and time consuming to develop.

Binford (1982), in reviewing current model-based research vision systems, concludes that most systems have not attempted to be general vision systems, although ACRONYM does demonstrate some progress toward this goal. Existing vision systems' performances are strongly limited by the performance of their segmentation modules, their weak use of world knowledge, and weak descriptions, making little use of shape.

With the exception of ACRONYM [and to an extent 3-D MOSAIC (Kanade, 1981)], the systems surveyed depend on image models and relations and therefore are strongly viewpoint dependent. To generalize to viewpoint-insensitive interpretations would require three-dimensional modeling and interpretations as in ACRONYM.

*Reprinted with permission from H. G. Barrow and J. M. Tenenbaum, "Computational Vision," *Proceedings of the IEEE*, Vol. 69, No. 5, May 1981. Copyright © 1981 by the Institute of Electrical and Electronics Engineers, Inc.

Binford concludes that although the results of these and other efforts are encouraging as first demonstrations, nevertheless as general vision systems, they have a long way to go.

8-5. PLAYERS

Rosenfeld, at the University of Maryland, issues a yearly bibliography, arranged by subject matter, related to the computer processing of pictorial information. The issue covering 1981 (Rosenfeld, 1982) includes nearly 1000 references.

The following is a list by category of the "principal players" in computer vision in the United States.

- *Research-oriented universities funded under DARPA* Image-Understanding Programs
  Carnegie-Mellon Univ.
  Univ. of Maryland
  MIT
  Univ. of Massachusetts
  Stanford Univ.
  Univ. of Rochester
  Univ. of Southern California
  Univ. of Rhode Island
- *Other active universities*
  Univ. of Texas, Austin
  Virginia Polytechnic Institute
  Purdue Univ.
  Univ. of Pennsylvania
  Univ. of Illinois
  Wayne State Univ.
  Johns Hopkins Univ.
  Rensselaer Polytechnic Institute
  Univ. of California, Berkeley
  North Carolina State Univ.
- *Nonprofit organizations*
  SRI International, AI Center
  JPL
  ERIM
- *U.S. government*
  NBS, Industrial Systems Division, Gaithersburg, MD
  NOSC (Naval Ocean Systems Center), San Diego
  NIH (National Institutes of Health)
- *Commercial vision systems developers:* Hundreds of companies are now involved in vision systems, a partial listing being given in Table 8-2.

TABLE 8-2 Commercial Vision System Developers

Industrial Vision Companies	Large Diversified Manufacturers[a]	Robot Manufacturers
Machine Intelligence Corp.	General Electric	Copperweld Robotics
Robot Vision Systems	Chrysler Corporation	Unimation
Videometrix	General Motors	Automatix, Inc.
Object Recognition Systems	Industrial Business Machines	
Octek, Inc.	Texas Instruments	
Cognex	Rockwell International	
Spectron Engineering, Inc.	Westinghouse	
Ham Industries	Hughes	
Quantomat	Lockheed–Palo Alto Research	
Image Recognition Systems	Lab.	
Colorado Video	Fairchild Camera and Instrument	
Everett Charles	Corp.	
Inspection Technology	Martin Marietta	
View Engineering	McDonald-Douglas Automation	
Vanzetti	Company	
Automated Vision Systems	Cheeseborough Ponds	
Perceptron, Inc.	Northrop	
Vicom Systems, Inc.	Minneapolis Honeywell	
Cyberanimation, Inc.	Boeing	
Reticon		
Control Automation		

[a] Some systems are for in-house use only.

8-6. RESEARCH TRENDS IN MODEL-BASED VISION SYSTEMS

Most research efforts in vision have been directed at exploring various aspects of vision, or toward generating particular processing modules for a step in the vision process rather than in devising general-purpose vision systems. However, there are currently two major U.S. efforts in general-purpose vision systems: the ACRONYM system at Stanford University under the leadership of T. Binford, and the VISIONS system at the University of Massachusetts at Amherst under A. Hanson and E. Riseman.

The ACRONYM system, outlined in Table 8-3A, is designed to be a general-purpose, model-based system that does its major reasoning at the level of volumes rather than images. The system basically takes a hierarchical top-down approach as in Fig. 8-5(B). ACRONYM has four essential parts: modeling, prediction, description, and interpretation. The user provides ACRONYM with models of objects (modeled in terms of volume primitives called *generalized cones*) and their spatial relationships; as well as generic models and their subclass relationships. These are both stored in graph form. The program automatically predicts which image features to expect. Description is a bottom-up process that generates a model-independent description of the image. Interpretation relates this description to the prediction to produce a three-dimensional understanding of the scene.

The VISIONS system outlined in Table 8-3B can be considered to be a work-

TABLE 8-3A Model-Based Vision Systems: ACRONYM

Developer: Brooks et al. (1979), Brooks (1981)
System: ACRONYM
Purpose: General-purpose vision system
Example domains: Identifying airplanes on a runway in aerial images
Simulation for robot systems and for automated grasping of objects

Approach	Modeling	Image Feature Extraction and Representation	Search and Matching	Remarks
Hierarchical top-down approach	Represents object classes from which subclasses and specific objects are represented by numeric constraints	Ribbons and curves obtained from an edge mapper	Matcher does an interpretation matching by mapping the observability graph into the picture graph	Aims to be a general vision system
Reasons between different levels of representation based on a hierarchy of representations		Surfaces obtained from a stereo mapper		Insensitive to viewpoint
High-level modeler provides a high-level language to manipulate models using symbolic names	Models three-dimensional objects using volume primitives: generalized cones and ribbons	Nodes of the picture graph (symbolic version of image) correspond to ribbons, surfaces, and curves. Arcs and relations indicate spatial relations between nodes	Matcher works in a coarse to fine order	A goal is to make use of total information for interpretation
Predictor and planner module is a rule-based system to generate an observability graph from the object graph (three-dimensional object representation consisting of nodes and relational arcs)	Spatial relations of volume elements within an object defined hierarchically		Combines local matches of ribbons into clusters	Feature extraction (e.g., finding lines and regions) still weak
Makes predictions (which are viewpoint insensitive) in the form of symbolic constraint expressions with variables	Can model both specific and generic volume elements and relations between them		Searches for maximal subgraph matches in the observability graph	Interpretation is limited to scenes with few objects
Makes a projective transformation from models	Models are part/whole graphs		Performs major interpretation at the level of volumes rather than at the level of images	Substantial progress has been achieved in past few years

Volume primitives have local rather than viewer-centered primitives

Predicts appearances of models in images in terms of ribbons and ellipses

Incorporates translation and rotation into observable representations

Searches for instances of models in images. It employs geometric reasoning in the form of a rule-based problem-solving system

It interprets (matches) in three dimensions by enforcing constraints of the three-dimensional model

TABLE 8-3B Model-Based Vision Systems: VISIONS

Developer: Hanson and Riseman (1978a, b)
Systems: VISIONS
Purpose: Interpreting static monocular scenes
Can be considered to be a working tool to test various image-understanding modules and approaches
Example domains: House scenes from ground level
Road scenes from ground level

Approach	Modeling	Image Feature Extraction and Representation	Search and Matching	Remarks
Uses hierarchical modular approach to representation and control	Hierarchical structure	Uses both edge finding and region growing to segment the image into a layered directed graph of regions, line segments, and vertices	Generates and stores partial models in "contexts" (of the CONNIVER programming language) which provide a history of decisions to be used when backtracking is necessary	System (Parma, 1980) did reasonably well in making a crude segmentation of a house scene
Tries to be as general as possible to allow both bottom-up and top-down solution hypotheses as well as various intermediate combinations	Scene schemas (like frames) are the highest representation	Uses a hierarchical processing cone (pyramid) to be able to handle image data at various levels of resolution		Viewpoint dependent
Incorporates the flexibility to utilize various feature extraction modules and multiple knowledge sources as required	Hierarchy is: Schemas Objects Volumes Surfaces	Uses a relaxation approach to organize edges into boundaries, and pixel clusters into regions, using high-level system guidance (interpretation guided segmentation)	Uses a multiple knowledge source heterarchical approach which generates partial models in the search space of models. At-	Schema used depends on specific scene
Allows for the possibility of generating and verifying hypotheses along many paths	Proposed representations of three-dimensional surfaces and volumes include Generalized cylinders Surface patches with			

tempts, using top-down and bottom-up relaxation techniques, to converge on a most probable solution

Uses rules for *focusing* on an element of a task, *expanding* that element by generating new hypotheses, and *verifying* new hypotheses

cubic B-splines to represent boundary and blending functions

Employs semantic networks

Nodes represent primitive entities (objects, concepts, situations, etc.)

Labeled arcs represent relationships between them

ing tool to test various image-understanding modules and approaches. Rather than using specific models, its high-level knowledge is in the form of framelike "schemas" which represent expectations and expected relationships in particular scene situations. VISIONS is based on monocular images and does its reasoning at the level of images rather than volumes.

Other research efforts in model-based vision systems are summarized in Tables III in Appendix I of Gevarter (1982a). All the research computer vision systems are individually crafted by the developers—reflecting the developers' backgrounds, interests, and domain requirements. All, except ACRONYM (and to an extent, 3-D MOSAIC), use image (two-dimensional) models and are viewpoint dependent. Models are described mostly by semantic networks, although feature vectors are also utilized. The systems, capitalizing on their choice to limit their observations to only a few objects, use predominantly the top-down interpretation-of-images approach, relying heavily on prediction.

8-7. FUTURE DIRECTIONS

As the field of computer vision unfolds, we expect to see the following future trends.

8-7.1. Techniques

- Although most industrial vision systems have used binary representations, we can expect increased use of gray scales because of their potential for handling scenes with cluttered backgrounds and uncontrolled lighting, and because the memory required for this more sophisticated approach is getting much less expensive.
- Recent theoretical work on monocular shape interpretation from images (shape from shading, texture, etc.) makes it appear promising that general mechanisms for generating spatial observations from images will be available before the end of the decade to support general vision systems.
- Successful techniques (such as stereo and motion parallax) for deriving shape and/or motion from multiple images should also be available during the last half of the 1980s.
- The mathematics of image understanding will continue to become more sophisticated.
- Enlargement will continue of the links now growing between image understanding and theories of human vision.

8-7.2. Hardware and Architecture

- We are now seeing hardware and software emerging that enables real-time operation in simple situations. Before the end of the decade, we should see hardware and software that will enable similar real-time operation for robotics and other activities requiring recognition, and position and orientation information.

- Fast raster-based pipeline preprocessing hardware to compute low-level features in local regions of an entire scene are now becoming available and should find general use in commercial vision systems within the next several years.

- As at virtually all visual levels, processing seems inherently parallel, parallel processing is a wave of the future (but not the entire answer).

- Relaxation and constraint analysis techniques are on the increase and will be increasingly reflected in future architectures.

8-7.3. AI and General Vision Systems

Computer vision will be a key factor in achieving many artificial intelligence applications. The goal is to move from special-purpose visual processing to general-purpose computer vision. Work to date in model-based systems has made a tentative beginning. But the long-run goal is to be able to deal with unfamiliar or unexpected input.* Reasoning in terms of generic models and reasoning by analogy are two approaches being pursued. However, it is anticipated that it will be a decade or more before substantial progress will be made.

8-7.4. Modeling and Programming

- Now emerging is three-dimensional modeling, arising largely from computer-aided design/computer-aided manufacturing (CAD/CAM) technology. Three-dimensional CAD/CAM data bases will be integrated with industrial vision systems to realistically generate synthesized images for matching with visual inputs.

- Illumination models, shading, and surface property models will be increasingly incorporated into visual systems.

- Volumetric models that allow prediction and interpretation at the levels of volumes, rather than images, will see greater utilization.

- High-level vision programming languages (such as Automatix's RAIL) that can be integrated with robot and industrial manufacturing languages are now beginning to appear and will become commonplace by the late 1980s.

- Generic representations for amorphous objects (such as trees) have been utilized experimentally and should become generally available by the end of the 1980s.

---

*As computer vision systems move toward this goal, they will increasingly incorporate expert system components using multiple knowledge sources. Chapter 6 and Gevarter (1982b) provides overviews of expert systems, in which ACRONYM and VISIONS are considered to be examples of expert systems.

8-7.5. Knowledge Acquisition

- Strategies for indexing into a large data base of models should be available within the latter half of this decade.
- "Training by being told" will supplement "training by example" as computer graphics techniques and vision programming languages become more common.

8-7.6. Sensing

- An important area of development is three-dimensional sensing. Several current industrial vision systems are already employing structured light for three-dimensional sensing. A number of new innovative techniques in this area are expected to appear in the next five years.
- More active vision sensors, such as scanning laser radars, are now being explored, and should find substantial industrial applications by the latter half of this decade.

8-7.7. Industrial Vision Systems

- We will see increased use of advanced vision techniques in industrial vision systems, including gray scale imagery.
- We are now observing a shortening time lag between research advances and their applications in industry. It is anticipated that in the future this lag may be as little as two years.
- Advanced electronics hardware at reduced cost is increasing the capabilities and speed of industrial vision while reducing costs.
- It is anticipated that special lighting and active sensing will play an increasing role in industrial vision.
- Common programming languages and improved interface standards will, within the next 3 to 10 years, enable easier integration of vision to robots and into the industrial environment.

8-7.8. Future Applications

- It is anticipated that about one-fourth of all industrial robots will be equipped with some form of vision system by 1990.
- It is likely that within the next decade on the order of half of all industrial inspection activities requiring vision will be done with computer vision systems.
- New vision system applications in a wide variety of areas, as yet unexplored, will begin to appear within this decade. An example of such a system might be visual traffic monitors at intersections that could perceive cars, pedestrians, and so on, in motion, and control the flow of traffic accordingly.

- Computer vision will play a large role in future military applications. For example, the Defense Mapping Agency intends to achieve fully automated production for mapping, charting, and geodesy by 1995, utilizing expert system-guided computer vision facilities.

In conclusion, the amount of activity and the many researchers in the computer vision field suggest that within the next 5 to 10 years we should see some startling advances in practical computer vision, although the availability of practical general vision systems still remains a long way off.

REFERENCES

Ballard, D. H., and **Brown, C. M.**, *Computer Vision*. Englewood Cliffs, NJ: Prentice-Hall, 1982.

Barrow, H. G., and **Tenenbaum, J. M.**, "Recovering Intrinsic Scene Characteristics from Images," in Hanson and Riseman, 1978, pp 3-26.

Barrow, H. G., and **Tenenbaum, J. M.**, "Computational Vision," *Proceedings of the IEEE*, Vol. 69, No. 5, May 1981, pp. 572-595.

Binford, T. O., "Survey of Model-Based Image Analysis Systems," *Robotics Research*, Vol. 1, No. 1, Spring 1982.

Brooks, R., "Symbolic Reasoning among 3-D Models and 2-D Images," *Artificial Intelligence*, Vol. 17, 1981, pp. 285-348.

Brooks, R., **Greiner, R.**, and **Binford, T. O.**, "The ACRONYM Model-Based Vision System," *Proceedings of the International Joint Conference on Artificial Intelligence 1979*, Vol. 6, pp. 105-113.

Clowes, M. B., "On Seeing Things," *Artificial Intelligence*, Vol. 2, 1971, pp. 79-116.

Cohen, P. R., and **Feigenbaum, E. A.**, Chap. 13, "Vision," *The Handbook of Artificial Intelligence*, Vol. 2. Los Altos, CA: W. Kaufmann, 1982, pp. 125-321.

Gevarter, W. B., "A Wiring Diagram of the Human Brain as a Model for Artificial Intelligence," *Proceedings of the IEEE International Conference on Cybernetics and Society*, Washington, DC, Sept. 1977, pp. 694-698.

Gevarter, W. B., *An Overview of Computer Vision*, NSIR 82-2582, National Bureau of Standards, Washington, DC, Sept. 1982a.

Gevarter, W. B., *An Overview of Expert Systems*, NBSIR 82-2505, National Bureau of Standards, Washington, DC, May 1982b.

Hanson, A. R., and **Riseman, E. M.** (Eds.), *Computer Vision Systems*. New York: Academic Press, 1978a.

Hanson, A. R., and **Riseman, E. M.** (1978b), "Segmentation of Natural Scenes," in Hanson and Riseman, 1978a, pp. 129-163.

Hiatt, B., "Toward Machines that See," *Mosaic*, Nov./Dec. 1981, pp. 2-8.

Huebell, D. H., and **Wessel, T. N.**, "Brain Mechanisms of Vision," *Scientific American*, Vol. 241, No. 3, Sept. 1979, pp. 150-163.

Huffman, D. A., "Impossible Objects as Nonsense Sentences," in *Machine Intelligence 6*, B. Meltzer and D. Michie (Eds.). New York: Wiley, 1971, pp. 295-323.

Ikeuchi, K., and Horn B. K. P., "Numerical Shape from Shading and Occluding Boundaries," *Artificial Intelligence*, 1981, 17.

Kanade, T., "Recovery of the Three-Dimensional Shape of an Object from a Single View," *Artificial Intelligence*, Vol. 17, Aug. 1981, pp. 409–460.

Marr, D. C., *Vision*. San Francisco: W. H. Freeman, 1982.

Marr, D., and Nishihara, H., "Visual Information Processing: Artificial Intelligence and the Sensorium of Sight," *Technology Review*, Oct. 1978, pp. 28-47.

Minsky, M. L., "A Framework for Representing Knowledge," in *The Psychology of Computer Vision*, P. H. Winston (Ed.). New York: McGraw-Hill, 1975, pp. 211-277.

Nevatia, R., *Machine Perception*. Englewood Cliffs, NJ, Prentice-Hall, 1982.

Parma, C. C., Hanson, A. M., and Riseman, E. M., "Experiments in Schema-Driven Interpretation of a Natural Scene," Univ. of Massachusetts, COINS Tech. Rep. 80-10, 1980.

Pratt, W. K., *Digital Image Processing*. New York: Wiley, 1978, pp. 568-587.

Reddy, R., and Newell, A., "Image Understanding: Potential Approaches," *ARPA Image Understanding Workshop*, Washington, DC, 1975.

Rosenfeld, A., "Picture Processing, 1981," Computer Vision Lab Rep. TR-1134, Univ. of Maryland, College Park, Jan. 1982. (Also in *Computer Graphics and Image Processing*, May 1982.)

Shairai, Y., "Analyzing Intensity Arrays Using Knowledge about Scenes," in Winston, 1975, pp. 93-114.

Tenenbaum, J. M., et al., "Prospects for Industrial Vision," in *Computer Vision and Sensor-Based Robots*, G. G. Dodd and L. Rossol (Eds.). New York: Plenum Press, 1979, pp. 239-259.

Waltz, D., "Understanding Line Drawings of Scenes with Shadows," in Winston, 1975, pp. 19-91.

Winston, P. H., (Ed.), *The Psychology of Computer Vision*. New York: McGraw-Hill, 1975.

9

NATURAL LANGUAGE PROCESSING*

9-1. INTRODUCTION

One major goal of artificial intelligence research has been to develop the means to interact with machines in natural language (in contrast to a computer language). The interaction may be typed, printed, or spoken. The complementary goal has been to understand how human beings communicate. The scientific endeavor aimed at achieving these goals has been referred to as computational linguistics (or more broadly as cognitive science), an effort at the intersection of AI, linguistics, philosophy, and psychology.

Human communication in natural language is an activity of the whole intellect, being associated with perception, thinking, and transmitting information, as well as reflecting human actions to achieve goals. AI researchers, in trying to formalize what is required to properly address natural language, find themselves involved in the long-term endeavor of having to come to grips with this entire activity. (Formal linguists tend to restrict themselves to the structure of language.) The current AI approach is to conceptualize language as a knowledge-based system for processing communications and to create computer programs to model that process.

Communication activities can serve many purposes, depending on the goals, intentions, and strategies of the communicator. The goal of the communication is to change some aspect of the recipient's mental state. Thus communication endeavors to add or modify information, change a mood, elicit a response, or establish a new goal for the recipient.

*A more complete treatment of natural language processing is given in Gevarter (1983).

For a computer program to interpret a relatively unrestricted natural language communication, a great deal of knowledge is required. Knowledge is needed of the following:

- The structure of sentences
- The meaning of words
- The morphology of words
- A model of the beliefs of the sender
- The rules of conversation
- An extensive shared body of general information about the world

This body of knowledge can enable a computer (like a human being) to use *expectation-driven processing*, in which knowledge about the usual properties of known objects, concepts, and what typically happens in situations can be used to understand incomplete or ungrammatical sentences in appropriate contexts.

There are many applications for computer-based natural language understanding systems. Some of these are listed in Table 9-1.

9-2. APPROACHES

Natural language processing (NLP) systems utilize both linguistic knowledge and domain knowledge to interpret the input. As domain knowledge (knowledge about the subject area of communication) is so important to understanding, it is usual to classify the various systems based on their representation and utilization of domain knowledge. On this basis, Hendrix and Sacerdoti (1981) classify systems as types A, B, or C,* with type A being the simplest, least capable, and correspondingly least costly systems.

9-2.1. Type A: No World Models

Keywords or patterns. The simplest systems utilize ad hoc data structures to store facts about a limited domain. Input sentences are scanned by the programs for predeclared keywords, or patterns, that indicate known objects or relationships. This is illustrated by ELIZA, an early keyword natural language interface system.

ELIZA (Weizenbaum, 1966) was the name of a family of programs that attempted to conduct conversations with human beings. ELIZA searched for keywords in a sentence and responded with one of several stored sentence patterns associated with that keyword. The program DOCTOR, which simulated a nondirective therapist, was the first example of an ELIZA script. A typical conversation is given by Figure B9-1.

*Other system classifications are possible (e.g., those based on the range of syntactic coverage).

TABLE 9-1   Some Applications of Natural Language Processing

Discourse

  Speech understanding

  Story understanding

Information access

  Information retrieval

  Question-answering systems

  Computer-aided instruction

Information acquisition or transformation

  Machine translation

  Document or text understanding

  Automatic paraphrasing

  Knowledge compilation

  Knowledge acquisition

Interaction with intelligent programs

  Expert systems interfaces

  Decision support systems

  Explanation modules for computer actions

  Interactive interfaces to computer programs

Interacting with machines

  Control of complex machines

Language generation

  Document or text generation

  Speech output

  Writing aids (e.g., grammar checking)

Men are all alike.

IN WHAT WAY

They're always bugging us about something or other.

CAN YOU THINK OF A SPECIFIC EXAMPLE

Well, my boyfriend made me come here.

YOUR BOYFRIEND MADE YOU COME HERE

He says I'm depressed much of the time.

I AM SORRY TO HEAR YOU ARE DEPRESSED

It's true, I am unhappy.

DO YOU THINK COMING HERE WILL HELP YOU NOT TO BE
  UNHAPPY

I need some help, that much seems certain.

WHAT WOULD IT MEAN TO YOU IF YOU GOT SOME HELP

Perhaps I could learn to get along with my mother.

TELL ME MORE ABOUT YOUR FAMILY

My mother takes care of me.

WHO ELSE IN YOUR FAMILY TAKES CARE OF YOU

My father.

YOUR FATHER

You are like my father in some ways.

WHAT RESEMBLANCE DO YOU SEE

You are not very aggressive but I think you don't want me to notice
  that.

WHAT MAKES YOU THINK I AM NOT VERY AGGRESSIVE

You don't argue with me.

WHY DO YOU THINK I DON'T ARGUE WITH YOU

You are afraid of me.

Figure B9-1. ELIZA; Example of an early keyword natural language processing system. Copyright 1966, Association for Computing Machinery, Inc., reprinted by permission.

DOES IT PLEASE YOU TO BELIEVE I AM AFRAID OF YOU
My father is afraid of everybody.
WHAT ELSE COMES TO MIND WHEN YOU THINK OF YOUR
    FATHER
Bullies.
DOES THAT HAVE ANYTHING TO DO WITH THE FACT THAT
    YOUR BOYFRIED MADE YOU COME HERE
The capitalized lines are the machine responses.*

Figure B9-1 (Continued)

The boxed words are typical of the keywords or phrases searched for by the program. The underlined words were transformed and were used to fill out the response patterns selected. If no keyword was spotted, the program would respond with a stock phrase such as "Please continue."

Limited logic systems.   In limited logic systems, information in their data base is stored in some formal notation, and language mechanisms are utilized to translate the input into the internal form. The internal form chosen is such as to facilitate performing logical inferences on information in the data base.

9-2.2. Type B: Systems That Use Explicit World Models

In these systems, knowledge about the domain is explicitly encoded, usually in frame or network representations (discussed previously in Section 4.2) that allow the system to understand input in terms of context and expectations. Cullingford's work (see Schank and Ableson, 1977) on SAM (Script Applier Mechanism) is a good example of this approach.

9-2.3. Type C: Systems That Include Information about the Goals and Beliefs of Intelligent Entities

These advanced systems (still in the research stage) attempt to include in their knowledge base information about the beliefs and intentions of the participants in the communication. If the goal of the communication is known, it is much easier to interpret the message. Schank and Abelson's (1977) work on plans and themes reflects this approach.

For more complex systems than those based on key words and pattern matching, language knowledge is required to interpret the sentences. The system usually begins by *parsing* the input (processing an input sentence to produce a more useful representation for further analysis). This representation is normally a structural description of the sentence indicating the relationship of the component parts. To address the parsing problem and to interpret the result, the computational linguistic community has studied syntax, semantics, and pragmatics. *Syntax* is the study of the structure of phrases and sentences. *Semantics* is the study of meaning. *Pragmatics* is the study of the use of language in context.

9-3. GRAMMARS

Barr and Feigenbaum (1981, p. 229) state: "A grammar of a language is a scheme for specifying the sentences allowed in the language, indicating the syntactic rules for combining words into well-formed phrases and clauses."* The following grammars are some of the most important.[†]

9-3.1. Phrase Structure Grammar: Context-Free Grammar

Chomsky (see, e.g., Winograd, 1983) had a major impact on linguistic research by devising a mathematical approach to language. He defined a series of grammars based on rules for rewriting sentences into their component parts. He designated these as 0, 1, 2, or 3, based on the restrictions associated with the rewrite rules, with 3 being the most restrictive.

Type 2—context-free or phrase structure grammar—has been one of the most useful in natural language processing. It has the advantage that all sentence structure derivations can be represented as a tree and practical parsing algorithms exist. Although it is a relatively natural grammar, it is unable to capture all the sentence constructions found in most natural languages, such as English. Gazder (1981) has recently broadened its applicability by adding augmentations to handle situations that do not fit the basic grammar. This generalized phrase structure grammar is now being developed by Hewlett-Packard (Gawron et al., 1982).

9-3.2. Transformational Grammar

Tennant (1981, p. 89) observes that "the goal of a language analysis program is recognizing grammatical sentences and representing them in a canonical structure (the underlying structure)." A *transformational grammar* (Chomsky, 1957) consists of a dictionary, a phrase structure grammar, and a set of transformations. In analyzing sentences, using a phrase structure grammar, first a parse tree (as in Fig. 9-3) is produced. This is called the *surface structure*. The transformational rules are then

---

*This and other excerpts from *The Handbook of Artificial Intelligence* are reprinted by permission of William Kaufmann, Inc., from A. Barr and E. A. Feigenbaum, *The Handbook of Artificial Intelligence*. Copyright 1981 by William Kaufmann, Inc., Los Altos, Calif. All rights reserved.

[†] Charniak and Wilks (1976) provide a good overview of the various approaches to grammars.

applied to the parse tree to transform it into a canonical form called the *deep* (or underlying) *structure*. As the same thing can be stated in several different ways, there may be many surface structures that translate into a single deep structure.

9-3.3. Case Grammar

Case grammar is a form of transformational grammar in which the deep structure is based on *cases*—semantically relevant syntactic relationships. The central idea is that the deep structure of a simple sentence consists of a verb and one or more noun phrases associated with the verb in a particular relationship. Fillmore (1971) proposed the following cases: agent, experiencer, instrument, object, source, goal, location, type, and path.

The cases for each verb form an ordered set referred to as a *case frame*. A case frame for the verb "open" would be

$$(object\ (instrument)\ (agent))$$

which indicates that open always has an object, but the instrument or agent can be omitted, as indicated by their surrounding parentheses. Thus the case frame associated with the verb provides a template which aids in understanding a sentence.

9-3.4. Semantic Grammars

For practical systems in limited domains, it is often more useful, instead of using conventional syntactic constituents such as noun phrases, verb phrases, and prepositions, to use meaningful semantic components instead. Thus, in place of nouns when dealing with a naval data base, one might use ships, captains, ports, and cargos. This approach gives direct access to the semantics of a sentence and substantially simplifies and shortens the processing. Grammars based on this approach are referred to as *semantic grammars* (see, e.g., Burton, 1976). An example of a system based on a semantic grammar is LADDER.

Actual conversations with the LADDER system are reproduced in Fig. B9-2. Characteristics of the LADDER system are given in Table B9-1.

9-3.5. Other Grammars

A variety of other, but less prominent grammars have been devised. Still others can be expected to be devised in the future. One example is Montague Grammar (Dowty et al., 1981), which uses a logical functional representation for the grammar and therefore is well suited for the parallel-processing logical approach now being pursued by the Japanese (see Nishida and Doshita, 1983) for their future AI work as embodied is their Fifth-Generation Computer research project.

TABLE B9-1 Characteristics of LADDER, a Research NLP System

Use	Approach	Capabilities	Limitations
Language access to distributed data with error recovery SRI (Hendrix et al., 1978) Natural language data base query	Application of LIFER parser Uses patterns or templates to interpret sentences; associates a function with each pattern Uses a semantic (pragmatic) grammar and associated functions to implicitly encode knowledge about language and the world; the grammar contains much information about the particular data base being queried Type A system	Can correct spelling Can handle ellipses Can interpret pronouns Can deal with large and complex data bases; e.g., in a naval ship data base has dealt with: 100 fields in 14 files Records of 40,000 ships Can answer certain questions based on its own natural language processing system Can be taught synonyms Can be taught new syntactic constructions Can accept a defined input sentence as equivalent to a whole set of questions	Conversation is limited strictly to questions about a small domain Cannot deal with logically complex notions such as: Disjunction Quantification Implication Causality Possibility Closed-world viewpoint; acts as if it were dealing with a world: Containing a fixed number of objects and relationships between them With objects and relationships being immutable

Boldface words represent system response. Material in square brackets replaces lengthy material not relevant to this illustration.

1—Give me the length of the Kenedy.
　　　　spelling-> KENNEDY

PARSED!
Give the length of the ship KENNEDY JF

LEN = 1072
2—width and draft
PARSED!
Trying Ellipsis: GIVE ME WIDTH AND DRAFT OF THE KENNEDY
Give the beam and draft of the ship KENNEDY JF
(BEAM 130 DFT 36)
3—Who is her commander?
Give the rank and name of the commander of the ship KENNEDY JF
(RANK CAPT NAME MOFFETT P)
4—What about the Fox?
PARSED!
Trying Ellipsis on: THE FOX
　　　　WHO IS THE FOX COMMANDER
Give the rank and name of the commander of the ship FOX
(RANK CAPT NAME EVERETT J)
5—What U.S. merchant ships carrying vanadium ore are within 500 miles of the Kimlow?
PARSED!

1—How far is the kennedy from the port of NY?
Trying Ellipsis: ELLIPSIS HAS FAILED
The parser does not expect the word "NY" to follow "HOW FAR IS THE KENNEDY FROM THE PORT OF"
Options for next word of symbol are: <PORT>
2—What is a <PORT>?
PARSED!
"<PORT>" may be any sequence of words following one of the patterns:

<PORT>	= >	CAPE TOWN
	= >	NEW YORK
	= >	NEW ORLEANS
		[etc.]

3—Define NY like New York.
PARSED!
NY
4—redo 1
[i.e., How far is the Kennedy from the port of NY?]
PARSED!
(GCDIST 1974)
5—Q1 Kennedy
[System gives syntax error.]
6—Let "Q1 kennedy" be a paraphrase of "what is the position, fuel status and homeport of the Kennedy"
PARSED!
<S> = > Q1 <SHIP>

Give the position of and time at position for the ship KIMLOW T
Give the name of merchant American ship S
    where the type of cargo on S is VNAD
    the great-circle distance from the position of S to
    15-33N, 30-10W, is less than or equal to 500
SHIP = GREENVILLE VICTORY, CRAIN ME, TOTOR
6—How fast are the U.S. subs with lengths greater than 150
PARSED!
Give the maximum cruising speed for and name of American ship S
    where the first character of the type of S is S
    the second character of the type of S is S
    the length of S is greater than 150 feet ?
[A long table relating the names and maximum cruising speeds of
American submarines is printed.]
7—What French ships are at readiness status 1?
PARSED!
Give the name of French ship S
    where the state of readiness of S is 1
There is no French ship S
NONE

7—q1 kennedy
PARSED!
(POSITION 4000N0600E PCFUEL 90 PORT MAYPORT)
8—American subs in the North Atlantic
Trying Ellipsis: Q1 AMERICAN SUBS IN THE NORTH ATLANTIC

SHIP	POSITION	PCFUEL	PORT
STURGEON	3700N7600W	100	NORFOLK
WHALE	3750N7700W	100	NORFOLK
ASPRO	3000N3000W	100	˙NORFOLK

[etc.]
9—Let "show the forcestatus of the Kitty Hawk" be like "Display the
    employment and readiness condition of the Kitty Hawk. Print her
    destination. List ships in her organization."
PARSED!
[New production added to system.]
10—show the forcestatus of Kennedy
PARSED!
[questions defined in 9 for Kitty Hawk are answered for the Kennedy.]
11—Define "Kennedy no nagasa wa ikura desuka" like "what is the
    length of the kennedy."
PARSED!
[Production added to system.]
12—Fox no nagasa wa ikura desuka?
PARSED!
LEN = 547

Figure B9-2 Conversations with the LADDER system. (From Hendrix and Sacerdoti, 1981, pp. 314, 348. Used by permission of Gary G. Hendrix and Earl D. Sacerdoti.)

9-4. SEMANTICS AND THE CANTANKEROUS ASPECTS OF LANGUAGE

Semantic processing (as it tries to interpret phrases and sentences) attaches meanings to the words. Unfortunately, English does not make this as simple as looking up the word in the dictionary, but provides many difficulties which require context and other knowledge to resolve. Examples are:

9-4.1. Multiple Word Senses

Syntactic analysis can resolve whether a word is used as a noun or a verb, but further analysis is required to select the sense (meaning) of the noun or verb that is actually used. For example, "fly" used as a noun may be a winged insect, a fancy fishhook, a baseball hit high in the air, or several other interpretations as well. The appropriate sense can be determined by context (e.g., for "fly" the appropriate domain of interest could be extermination, fishing, or sports), or by matching each noun sense with the senses of other words in the sentence. The latter approach was taken by Reiger and Small (1979) using the (still embryonic) technique of "interacting word experts," and by Finin (1980) and McDonald (1980) as the basis for understanding noun compounds.

9-4.2. Pronouns

Pronouns allow a simplified reference to nouns, sets, or events previously used (or implied). Where feasible, using pragmatics, pronoun antecedents are usually identified by reference to the most recent noun phrase having the same context as the pronoun.

9-4.3. Ellipsis and Substitution

Ellipsis is the phenomenon of not stating explicitly some words in a sentence, but leaving it to the reader or listener to fill them in. (Observe the way ellipsis is handled in the LADDER example.) Substitution is similar—using a dummy word in place of the omitted words. Employing pragmatics, ellipses and substitutions are usually resolved by matching the incomplete statement to the structures of previous recent sentences—finding the best partial match and then filling in the rest from this matching previous structure.

9-5. SYNTACTIC PARSING

Parsing assigns structures to sentences. The following types have been developed over the years for NLP (Barr and Feigenbaum, 1981).

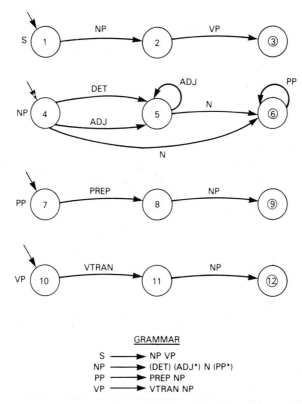

GRAMMAR

$$S \longrightarrow NP\ VP$$
$$NP \longrightarrow (DET)\ (ADJ^*)\ N\ (PP^*)$$
$$PP \longrightarrow PREP\ NP$$
$$VP \longrightarrow VTRAN\ NP$$

Figure 9-1 Transition network for a small subset of English. Each diagram represents a rule for finding the corresponding word pattern. Each rule can call on other rules to find needed patterns. (From Graham, 1979, p. 214. Used by permission of TAB Books, Blue Ridge Summit, PA.)

9-5.1. Template Matching

Most of the early (and some current) NLP programs performed parsing by matching their input sentences against a series of stored templates.

9-5.2. Transition Nets

Phrase structure grammars can be syntactically decomposed using a set of rewrite rules such as those indicated in Fig. 9-1. Observe that a simple sentence can be rewritten as a noun phrase and a verb phrase as indicated by

$$S \rightarrow NP\ VP$$

The noun phrase can be rewritten by the rule

$$NP \rightarrow (DET)(ADJ^*)N(PP^*)$$

where the parentheses indicate that the item is optional, and the asterisks (associated with the adjectives and prepositional phrases) indicate that any number of items may occur. An example of an analyzed noun phrase is shown in Figs. 9-2 and 9-3.

As the transition networks analyze a sentence, they can collect information about the word patterns they recognize and fill slots in a frame associated with each pattern. Thus they can identify noun phrases as singular or plural, whether the nouns refer to persons and if so their gender, and so on, needed to produce a deep structure. A simple approach to collecting this information is to attach subroutines to be called for each transition. A transition network with such subroutines attached is called an *augmented transition network* (ATN). With ATNs, word patterns can be recognized. For each word pattern, we can fill slots in a frame. The resulting filled frames provide a basis for further processing.

9-5.3. Other Parsers

Other parsing approaches have been devised, but ATNs remain the most popular syntactic parsers. ATNs are *top-down parsers* in that the parsing is directed by an anticipated sentence structure. An alternative approach is *bottom-up parsing*, which examines the input words along the string from left to right, building up all possible structures to the left of the current word as the parser advances. A bottom-up parser could thus build many partial sentence structures that are never used, but the diversity could be an advantage in trying to interpret input word strings that are not clearly delineated sentences or contain ungrammatical constructions or unknown words. There have been recent attempts to combine the top-

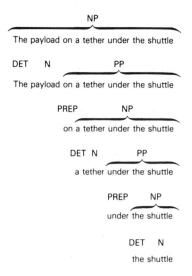

Figure 9-2 Example noun phrase decomposition.

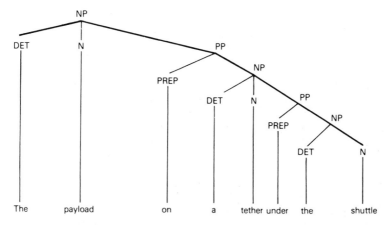

Figure 9-3 Parse tree representation of the noun phrase surface structure.

down with the bottom-up approach for NLP in a manner similar to that used for computer vision.

9-6. SEMANTICS, PARSING, AND UNDERSTANDING

The role of syntactic parsing is to construct a parse tree or similar structure of the sentence to indicate the grammatical use of the words and how they are related to each other. The role of the semantic processing is to establish the meaning of the sentence. This requires facing up to all the cantankerous ambiguities discussed earlier.

Charniak (1981) observes that there have been two main lines of attack on word sense ambiguity. One is the use of discrimination nets (Reiger and Small, 1979) that utilize the syntactic parse tree (by observing the grammatical role that the word plays, such as taking a direct object, etc.) in helping to decide the word sense. The other approach is based on the frame/script idea (used, e.g., for story comprehension) that provides a context and the expected sense of the word (see, e.g., Schank and Abelson, 1977).

Charniak indicates that the semantics at the level of the word sense is not the end of the parsing process, but what is desired is understanding or comprehension (associated with pragmatics). Here the use of frames, scripts, and more advanced topics such as plans, goals, and knowledge structures (see, e.g., Schank and Riesbeck, 1981) play an important role.

9-7. NATURAL LANGUAGE PROCESSING SYSTEMS

As indicated below, various NLP systems have been developed for a variety of functions.

9-7.1. Kinds of Systems

Question-answering systems.   Question-answering natural language systems have perhaps been the most popular of the NLP research systems. They have the advantage that they usually utilize a data base for a limited domain and that most of the user discourse is limited to questions.

Natural language interfaces (NLIs).   These systems are designed to provide a painless means of communicating questions or instructions to a complex computer program.

Computer-aided instruction (CAI).   Arden (1980, p. 465) states:

> One type of interaction that calls for ability in natural languages is the inter-action needed for effective teaching machines. Advocates of computer-aided instruction have embraced numerous schemes for putting the computer to use directly in the educational process. It has long been recognized that the ulti-mate effectiveness of teaching machines is linked to the amount of intelligence embodied in the programs. That is, a more intelligent program would be better able to formulate the questions and presentations that are most appro-priate at a given point in a teaching dialogue, and it would be better equipped to understand a student's response, even to analyze and model the knowledge of the student, in order to tailor the teaching to his needs.*

Discourse.   Systems that are designed to understand discourse (extended dialogue) usually employ pragmatics. Pragmatic analysis requires a model of the mutual beliefs and knowledge held by the speaker and listener.

Text understanding.   Although Schank (see Schank and Riesbeck, 1981) and others have addressed themselves to this problem, much more remains to be done. Techniques for understanding printed text include scripts and causative ap-proaches. Text understanding is important not only in deciphering a message, but as an important step in language translation.

Text generation.   There are two major aspects of text generation: one is the determination of the content and textual shape of the message, the second is trans-forming it into natural language. There are two approaches for accomplishing this. The first is indexing into canned text and combining it as appropriate. The second is generating the text from basic considerations. McDonald's thesis (1980) provides one of the most sophisticated approaches to text generation.

*Reprinted by permission from B. W. Arden, ed., *What Can Be Automated?* Copyright 1980 by MIT Press, Cambridge, Mass.

9-7.2 Research NLP Systems

Until recently, virtually all the NLP systems generated were of a research nature. These NLP systems basically were aimed at serving five functions:

1. Interfaces to computer programs
2. Data base retrieval
3. Text understanding
4. Text generation
5. Machine translation

Gevarter (1983) includes a survey of research NLP systems.

9-7.3. Commercial Systems

The commercial systems available in 1984 (together with their approximate price) are listed in Table 9-2. Several of these systems are derivatives of past research NLP systems.

9-8. CURRENT STATE OF THE ART

It is now feasible to use computers to deal with natural language input in highly restricted contexts. However, interacting with people in a facile manner is still far off, requiring understanding of where people are coming from—their knowledge, goals, and moods.

In today's computing environment, the only systems that perform robustly and efficiently are type A systems—those that do not use explicit world models, but depend on keyword or pattern matching and/or semantic grammars. In actual working systems, both understanding and text generation, ATN-like grammars can be considered the state of the art, although the use of rule-based expert systems appears to be emerging.

9-9. PLAYERS AND RESEARCH TRENDS

9-9.1. Players

Because of the growing interest in NLP, many universities and a number of industrial firms are involved in NLP research and development. Among the most active are Yale University, the University of California at Berkeley, Carnegie-Mellon University, MIT, the University of Illinois, SRI, MITRE, IBM, Texas Instruments, Artificial Intelligence Corporation, and Cognitive Systems, Inc.

TABLE 9-2 Some Commercial Natural Language Systems

System	Organization	Purpose	Comments
INTELLECT (derivative of ROBOT); $50,000/system; also distributed as ON-LINE ENGLISH (Culliane) and GRS Executive (Information Sciences)	Artificial Intelligence Corp., Waltham, MA	NLI for data base retrieval (other extensions under way)	Several hundred systems sold Takes about two weeks to implement for a new data base Written in PL-1 Available for mainframes Now available through IBM
PEARL (based on SAM and PAM); $250,000/system	Cognitive Systems, New Haven, CT	Custom NLIs: the first system—Explorer—is an interface to an existing map-generating system; others are interfaces to data bases	Large startup cost in building the knowledge base Several systems have been, and are being, built Written in LISP
NLI for personal computers (derivative of LIFER); less than $1000	Symantec, Cupertino, CA	Highly portable NLI for Data Base Management Systems for microcomputers	Written in Pascal; designed to be very compact and efficient; available early 1985 User customized
SAVVY; $349–$950	Excalibur Technologies Corp., Albuquerque, NM	System interface for microcomputers	Not linguistic; uses adaptive (best fit) pattern matching to strings of characters Released 3/82 User customized
NLI to R:Base; $195	Microrim, Inc., Bellvue, WA	System interface to a data base management system for personal computers	An expert system that uses transition networks and some 500 rules to parse and interpret the input Can handle multirelation queries
NLMENU	Texas Instruments, Inc., Dallas, TX	NLI to relational data bases microcomputers	Menu-driven NL query system All queries constructed from menu fall within linguistic and conceptual coverage of the system; therefore, all queries entered are successful

			Grammars used are semantic grammars written in a context-free grammar formalism
			Producing an interface to any arbitrary set of relations is automated and requires only a 15 to 30-minute interaction with someone knowledgeable about the relations in question
			$150 NLI to Dow Jones News/ Retrieval Service available
			Natural link tool kit available for original-equipment manufacturers to build their own NLIs
Weidner System; $16,000/ language direction	Weidner Communications Corp, Provo, UT	Semiautomatic natural language translation	Linguistic approach; written in FORTRAN IV
			Translation with human editing is approximately 100 words per hour (up to eight times as fast as human being alone)
			Approximately 20 sold by end of 1982, mainly to large multinational corporations
ALPS	ALPS, Provo, UT	Interactive natural language translation	Linguistic approach
			Uses a dictionary that provides the various translations for technical words as a display to human translator, who then selects among the displayed words

9-9.2. Research Trends

Current research in natural language processing systems includes machine translation, information retrieval, and interactive interfaces to computer systems. Important supporting research topics are language and text analysis, user modeling, domain modeling, task modeling, discourse modeling, reasoning, and knowledge representation.

Much of the research required (as well as the research now under way) is centered around addressing the problems and issues in the following areas:

How people use language.   The psychological mechanisms underlying human language production provide a fertile field for investigation. Efforts are needed to build explicit computational models to help explain why human languages are the way they are and the role they play in human perception.

Linguistics.   Further research is needed on methods for disambiguating language and for the utilization of context in language understanding.

Conversation.   Additional work is needed on ways to represent the huge amount of knowledge needed for natural language understanding. A great deal of research is needed to give understanding systems the ability to understand not only what is actually said, but the underlying intention as well.

Research is now under way by many groups on explicitly modeling goals, intentions, and planning abilities of people. Investigation of script and frame-based systems is currently the most active natural language understanding research area.

NLP system design.   Architectures, grammars, parsing techniques, and internal representations needed for NLP systems remain important research areas. One particularly fertile area is how best to utilize semantics to guide the path of the syntactic parser. Charniak (1981, p. 1085) indicates that a relatively unexplored area requiring research is the interaction between the processes of language comprehension and the form of semantic representation used.

Further work is needed on bringing multiple knowledge sources (KSs: syntactic, semantic, pragmatic, and contextual) to bear on understanding a natural language utterance, but still keeping the KSs separate for easy updating and modification. Also needed is further work in AI problem solving to cope with the problem of finding an appropriate structure in the huge space of possible meanings of a natural language input.

Improved NL understanding techniques are needed to handle complex notions such as disjunction, quantification, implication, causality, and possibility. Also needed are better methods for handling "open worlds," where all things needed to understand the world are not in the system's knowledge base. Further research is also necessary to aid with a common source of trouble in NLP: dealing with syntactic and semantic ambiguities and how to handle metaphors and idioms. Finally, the problems of efficiency, speed, portability, and so on, are all in need of better solutions.

Data base interfaces.   A current research topic is how data base schemas can best be enriched to support a natural language interface, and what the best logical structure would be for a particular data base. Research is also needed on more efficient methods for compiling a vocabulary for a particular application.

Text understanding.   Seeking general methods of concept extraction remains as one of the major research areas in text understanding.

9-10. FUTURE DIRECTIONS

Commercial natural language interfaces (NLIs) to computer programs and data base management systems are now becoming available. The advent of NLIs for microcomputers is the precursor to eventually making it possible for virtually anyone to have direct access to powerful computational systems.

As the cost of computing hardware has continued to fall, but the cost of programming has not, it has already become cheaper in some applications to create NLI systems (that utilize subsets of English) than to train people in formal programming languages.

Computational linguists and workers in related fields are devoting considerable attention to the problems of NLP systems that understand the goals and beliefs of the individual communicators. Although progress has been made and feasibility has been demonstrated, more than a decade will be required before useful systems with these capabilities become available.

One of the problems in implementing new installations of NLP systems is gathering information about the applicable vocabulary and the logical structure of the associated data bases. Work is now under way to develop tools to help automate this task. Such tools should be available by the end of the decade.

For text understanding, experimental programs have been developed that "skim" stylized text such as short disaster stories in newspapers (DeJong, 1982) to extract the key points. Despite the practical problems of sufficient world knowledge and the extension of language required, practical tools emerging from these efforts should be available to provide assistance to human beings doing text understanding within this decade.

The NRL Computational Linguistic Workshop (1981) concluded that text generation techniques are maturing rapidly and that new application possibilities will appear within the next five years.

The NRL workshop also indicated that:

Machine aids for human translators appear to have a brighter prospect for immediate application than fully automatic translation; however, the Canadian

French-English weather bulletin project is a fully automatic system in which only 20% of the translated sentences require minor rewording before public release. An ambitious common market project involving machine translation among six European languages is scheduled to begin shortly. Sixty people will be involved in that undertaking which will be one of the largest projects undertaken in computational linguistics.* The panel was divided in its forecast on the five year perspective of machine translation but the majority were very optimistic.

Nippon Telegram and Telephone Corp. in Tokyo has a machine translation AI project under way. An experimental system for translating from Japanese to English, and vice versa, has been demonstrated. In addition, the Japanese Fifth-Generation Computer effort has computer-based natural language understanding as one of its major goals.

In summary, natural language interfaces using a limited subset of English are now becoming available. Hundreds of specialized systems are already in operation. Major efforts in text understanding and machine translation are under way, and useful (though limited) systems are emerging. Systems that are heavily knowledge-based and handle more complete sets of English should be available within this decade. However, systems that can handle unrestricted natural discourse and understand the motivation of the communicators remain a distant goal, probably requiring more than a decade before useful systems appear.

REFERENCES

Arden, B. W. (Ed.), *What Can Be Automated?* Cambridge, MA: MIT Press, 1980.

Barr, A., and Feigenbaum, E. A., Chap. 4, "Understanding Natural Language," *The Handbook of Artificial Intelligence.* Los Altos, CA: W. Kaufmann, 1981, pp. 233–321.

Burton, R. R., "Semantic Grammar: An Engineering Technique for Constructing Natural Language Understanding Systems," BBN Rep. 3453, Bolt, Beranek and Newman, Cambridge, MA, Dec. 1976.

Charniak, E., "Six Topics in Search of a Parser: An Overview of AI Language Research," *IJCAI-81*, pp. 1079–1087.

Charniak, E., and Wilks, Y., *Computational Semantics.* Amsterdam: North-Holland, 1976.

Chomsky, N., *Syntactic Structures.* The Hague: Mouton, 1957.

DeJong, G., "An Overview of the FRUMP System," in *Strategies for Natural Language Processing*, W. G. Lehnert and M. H. Ringle (Eds.). Hillsdale, NJ: Lawrence Erlbaum, 1982, pp. 149–176.

Dowty, R., et al., *Introduction to Montague Semantics.* Boston: D. Reidel, 1981.

*EUROTA—a machine translation project sponsored by the European Common Market—eight countries, more than 15 universities, $24 million over several years.

Fillmore, C., "Some Problems for Case Grammar," in *Report of the Twenty-Second Annual Round Table Meeting on Linguistics and Languages Studies*, R. J. O'Brien (Ed.). Washington, DC: Georgetown Univ. Press, 1971, pp. 35–56.

Finin, T. W., "The Semantic Interpretation of Compound Nominals," Ph.D. thesis, Univ. of Illinois, Urbana, 1980.

Gawron, J. M., et al., "Processing English with a Generalized Phrase Structure Grammar," *Proceedings of the 20th Meeting of ACL*, Univ. of Toronto, Canada, June 16–17, 1982, pp. 74–81.

Gazdar, G., "Unbounded Dependencies and Coordinate Structure," *Linguistic Inquiry*, Vol. 12, 1981, pp. 155–184.

Gevarter, W. B., *An Overview of Computer-Based Natural Language Processing*, NBSIR 83-2687, National Bureau of Standards, Washington, DC, Apr. 1983 (also NASA TM 85635).

Graham, N., *Artificial Intelligence*. Blue Ridge Summit, PA: TAB Books, 1979.

Hendrix, G. G., and Sacerdoti, E. D., "Natural-Language Processing: The Field in Perspective," *Byte*, Sept. 1981, pp. 304–352.

Hendrix, G. G., Sacerdoti, E. D., Sagalowicz, D., and Slocum, J., "Developing a Natural Language Interface to Complex Data," ACM *Transactions on Database Systems*, Vol. 3, No. 2, June 1978.

McDonald, D. D., "Natural Language Production as a Process of Decision-Making under Constraints," Ph.D thesis, MIT, Cambridge, MA, 1980.

Nishida, T., and Doshita, S., "An Application of Montague Grammar to English–Japanese Machine Translation," *Proceedings of the Conference on Applied Natural Language Processing*, Santa Monica, CA, Feb. 1983.

NRL Computational Linguistic Workshop, "Applied Computational Linguistics in Perspective," Stanford University, June 26–27, 1981. (Proceedings in *American Journal of Computational Linguistics*, Vol. 8, No. 2, Apr.–June 1982, pp. 55–83.)

Reiger, C., and Small, S., "Word Expert Parsing," *Proceedings of the Sixth International Joint Conference on Artificial Intelligence*, 1979, pp. 723–728.

Schank, R. C., and Abelson, R. P., *Scripts, Plans, Goals and Understanding*. Hillsdale, NJ: Lawrence Erlbaum, 1977.

Schank, R. C., and Riesbeck, C. K., *Inside Computer Understanding*. Hillsdale, NJ: Lawrence Erlbaum, 1981.

Tennant, H., *Natural Language Processing*. New York: Petrocelli Books, 1981.

Weizenbaum, J., "Eliza–A Computer Program for the Study of Natural Language Communication between Man and Machine," *Communications of the ACM*, Vol. 9, No. 1, 1966, pp. 36–45. (Reprinted in *Communications of the ACM*, Vol. 26, No. 1, Jan. 1983, pp. 23–27.)

Winograd, T., *Language as a Cognitive Process*, Vol. I: Syntax. Reading, MA: Addison-Wesley, 1983.

10

SPEECH RECOGNITION AND SPEECH UNDERSTANDING*

10-1. INTRODUCTION

Speech is our fastest means of discourse communication, being about twice as fast as the average typist. It is also nearly effortless: speech does not need visual or physical contact and it places few restrictions on the use of the hands or the mobility of the body. Speech is thus well suited to communication with a machine when a person is engaged in other activities. Its effortlessness also makes it desirable for operating a computer, and it is a long-term candidate for direct text preparation (automatic dictation).

There are many applications emerging for speech recognition and speech understanding systems. Some of these are listed in Tables 10-1 and 10-2. Speech understanding systems have all the difficulties of natural language understanding plus the problem of interpreting the speech signal with all its noise and variability. As a result, speech understanding is one of the most difficult AI subjects, being a perception task related to the scene understanding problem in computer vision. Although the constraining aspects of natural language help reduce the magnitude of the task, it remains a major problem area.

Speech systems can be categorized into speech recognition systems and speech understanding systems, the former task being considerably easier. In addition, the systems divide further into those that work with isolated words and those that can handle connected speech, the latter being perhaps an order of magnitude more difficult than the former. Finally, speech systems are also classified as speaker dependent and speaker independent. The former systems are trained to recognize the

*An overview of speech synthesis is given in Appendix F.

TABLE 10-1   Speech Recognition Applications

Manufacturing processes and control
  Quality control data entry into computers
  Shipping and receiving—record entry, package sorting
  Maintenance and repair orders—part availability, work needed or under way
  CAD/CAM (Computer-aided design/computer-aided manufacturing)

Office automation
  Executive workstation
  Word processing
  Data entry
  Control functions

Technical data gathering
  Cartography—inputs when working with maps
  Working with blueprints
  Medical applications
    Dental records
    Pathology
    Services for the handicapped
    Operating room logging
    Command/control of medical instrumentation

Security applications
  Building access
  Computer file access
  Communications security
  Speaker verification/identification

Consumer products applications
  Control functions
  Status queries

Equipment subsystem operation
  Aircraft
  Spacecraft
  Military equipment

---

particular speakers using it. The heart of the speech problem (that gives rise to the above classifications) is the difficulty of recognizing the speech signal.

10-2. THE NATURE OF SPEECH SOUNDS

Acoustics and phonetics may be the key to speech understanding. Zue (1981) argues that human spectrograph-reading experiments indicate that phonetic recognition in speech systems can be improved substantially, which would result in much more capable speech systems.

Speech recognition is based primarily on the identification of words. An adult speaker may know 100,000 of the 300,000 words in the English language. Each language has a basic set of speech sounds called *phonemes*. In English there are only about 40 phonemes, compared with some 10,000 for the next largest speech unit, the *syllable*.

TABLE 10-2 Speech Understanding Applications

Universal access to large data bases via the telephone network

Automatic telephone transaction systems: airline reservations and inquiries

Command and control
   Military
   Business

Operation of complex machines

The sounds that make up human speech are generated by the flow of air through the vocal tract in three ways (Levinson and Liberman, 1981):

1. The vocal cords can be made to vibrate, resulting in the frequency of the sound referred to as *pitch*.
2. A constriction can be formed in the vocal tract, narrow enough to cause turbulence, resulting in noise-like sounds such as that used to produce f.
3. Pressure built up behind a closure (such as the lips) can release a burst of acoustic energy, as in the pronunciation of consonants, such as p, t, and k.

These three sources of speech sound are shaped acoustically by the time-varying physical shape of the vocal tract.

One way to characterize the speech signal is by its Fourier transform, which specifies the amplitude and phase of each of the frequencies in the frequency-domain representation (spectrum) of the signal. As the phase makes little perceptual difference, the signal is represented in practice by its amplitude spectrum, in a representation called a *spectrograph*.

10-3. SPEECH RECOGNITION

10-3.1. Isolated Word Recognition

Figure 10-1 indicates a basic paradigm for speech recognition. The signal is first operated on to emphasize the frequency range 2 to 3 kHz, filtered to chop off high frequencies (>8 kHz), then digitized. The end points of the word are detected, and a set of parameters representing the word are generated. This is then matched with stored parameter sets in the system's vocabulary, and the word with the closest match is chosen. For a word, the acoustic signal varies both in duration and amplitude each time the same speaker says it. Thus it may have to be warped to achieve the best comparison with the reference, this task being one of the toughest problems for a speech recognizer. The warping is usually accomplished by dynamic programming.

Doddington and Schalk (1981, p. 28) state that:

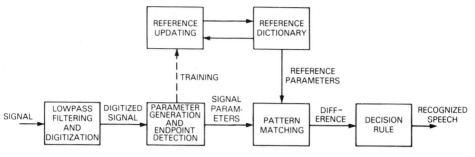

After Zue (1982, p51)

Figure 10-1 Basic speech recognition paradigm. (From Zue, 1982, p. 51. Used by permission of D. Reidel Publishing Company, Dordvecht, Holland.)

The most common means of feature extraction is direct measurement of spectrum amplitude, with, for example, a set of 16 bandpass filters. Another means is measurement of the zero-crossing rate of the signal in several broad frequency bands to give an estimate of the formant [resonant] frequencies in these bands. Yet another means is representing the speech signal in terms of the parameters of a filter whose spectrum best fits that of the input speech signal. This technique known as linear predictive coding (LPC) has gained popularity because it is efficient, accurate, and simple.

10-3.2. Recognizing Continuous Speech

For continuous speech, rather than attempting to match all possible word patterns, it is often more efficient to work with speech units much smaller than words, particularly phonemes. Breaking down the speech signal into these smaller components and giving them symbols is referred to as *segmentation and labeling*. Usually, several phoneme labels are assigned to each segment by a pattern-matching process, which also assigns a probability value representing the goodness of the match. With the appropriate acoustic-phonetic knowledge, it is possible to combine, regroup, and delete segments to form larger phoneme units. The lexical knowledge of word pronunciations can now be used to generate a multiplicity of word hypotheses. For a sufficiently limited vocabulary, and perhaps also employing some syntactic and word boundary knowledge, speech recognition can be achieved.

10-4. SPEECH UNDERSTANDING

Arden (1980, pp. 475, 478) observes that:

Speech-understanding systems differ somewhat from recognition systems, in that they have access to and make effective use of task-specific knowledge in the analysis and interpretation of speech. Further, the criteria for performance are somewhat relaxed, in that the errors that count are not the errors in speech recognition, but errors in task accomplishment.

To successfully decode the unknown utterance, a speech perception system must effectively use the many diverse sources of knowledge about the language, the environment, and the context. These sources of knowledge include the characteristics of speech sounds (acoustic-phonetic), variability in pronunciation (phonology), the stress and intonation patterns of speech (prosodics), the sound patterns of words and sentences (lexicon), the grammatical structure of language (syntax), the meaning of words and sentences (semantics), and the context of the conversation (pragmatics). . . .

What makes speech perception a challenging and difficult area of A.I. is the fact that error and ambiguity permeate all the levels of the speech-decoding process. . . .

The grammatical structure of sentences can be viewed principally as a mechanism for reducing search by restricting the number of acceptable alternatives.*

Barr and Feigenbaum (1981, p. 332) note that the types of knowledge at the various levels in processing spoken knowledge include (from the signal level up):

1. *Phonetics:* representations of the physical characteristics of the sounds in all of the words in the vocabulary.

2. *Phonemics:* rules describing variations in pronunciation that appear when words are spoken together in sentences (coarticulation across word boundaries, "swallowing" of syllables, etc.).

3. *Morphemics:* rules describing how morphemes (units of meaning) are combined to form words (formation of plurals, conjugations of verbs, etc.).

4. *Prosodics:* rules describing fluctuation in stress and intonation across a sentence.

5. *Syntax:* the grammar or rules of sentence formation resulting in important constraints on the number of sentences (not all combinations of words in the vocabulary are legal sentences).

6. *Semantics:* the "meaning" of words and sentences, which can also be viewed as a constraint on the speech understander (not all grammatically legal sentences have a meaning—e.g., The snow was loud).

7. *Pragmatics:* rules of conversation (in a dialogue, a speaker's response must not only be a meaningful sentence but also be a reasonble reply to what was said to him). For instance, it is pragmatic knowledge that tells us that the question "Can you tell me what time it is?" requires more than just a Yes or No response.

Using this knowledge, the hierarchical structure leading to speech understanding can be characterized as shown in Fig. 10-2.

---

*Reprinted by permission from B. W. Arden, ed., *What Can Be Automated?* Copyright 1980 by MIT Press, Cambridge, Mass.

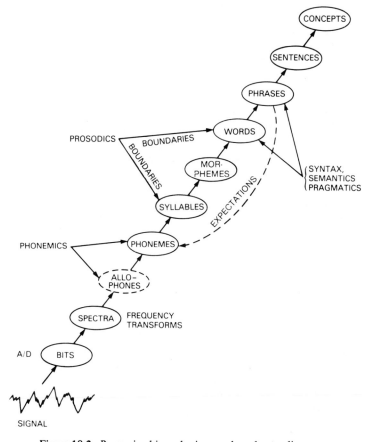

Figure 10-2   Processing hierarchy in speech understanding.

10-5. THE ARPA SPEECH UNDERSTANDING RESEARCH PROJECT

10-5.1. Introduction

In 1971, ARPA (the Advanced Research Projects Agency, now **DARPA**) initiated a five-year speech understanding research effort that proved to be one of the most significant projects in AI history. Not only did it greatly advance our knowledge of speech, but it provided new insights on how to structure and control a complex expert system.

Lea and Shoup (1979) reported that the **ARPA** Speech Understanding Research (SUR) project had the highly ambitious goals of understanding, with 90% accuracy, continuous speech from a 1000-word vocabulary spoken by several cooperative speakers under near-ideal conditions of quiet rooms and high-fidelity equipment. It was intended that the processing take no more than several times real time using large, very fast computers.

There were three principal complete systems developed under the project: Hearsay II and HARPY at Carnegie-Mellon University (CMU) and HWIM (Hear What I Mean) at Bolt, Beranek and Newman (BBN). In 1976, the ARPA goals were essentially met at CMU by HARPY exhibiting 95% accuracy and Hearsay II achieving 90% accuracy. HWIM had a substantially lower accuracy, but utilized a more difficult vocabulary. (HWIM's domain was travel budget management. Hearsay II's and HARPY's was retrieval of AI documents.) These three systems were heavily knowledge-based and are now considered to be expert systems.

All the ARPA SUR systems utilized a combination of bottom-up and top-down processing. The lower levels used knowledge about the variable phonetic composition of the words in the vocabulary (lexicon) to interpret pieces of the speech signal by comparing it with prestored patterns. The top level aided in recognition by building expectations about which words the speaker was likely to say, using syntactic and semantic constraints (Barr and Feigenbaum, 1982, pp. 326–327).

10-5.2. Hearsay II

Hearsay II is characterized by its cooperative problem-solving system architecture (see Fig. 10-3) which employs a set of programmed "specialists" (knowledge sources: KSs) interacting via a shared common "blackboard" on which their decisions were recorded. The blackboard can be visualized as a global data structure representing a multilevel network of alternative hypotheses.

Hearsay has a total of 12 KSs, which at the lower levels created syllable class hypotheses from segments, word hypotheses from syllables, and so on. At the higher levels, KSs acted to predict all possible words that might syntactically precede or follow a phrase, create phrase hypotheses from verified contiguous word-phrase pairs, and so on.

The majority of the hypotheses contributed by the KSs at any level did not end up in the final interpretation of the sentence. Instead, only the most likely hypotheses were chosen for expansion. The individual KSs operated somewhat independently and asynchronously through pattern-invoked programs when matching patterns appeared on the blackboard. To economize on computing resources, each hypothesis was rated and (using an appropriate scheduling routine) the most likely patterns were expanded first.

10-5.3. HARPY

A crude way of thinking of HARPY is as a compiled version of Hearsay II. HARPY uses a single precompiled network knowledge structure. Barr and Feigenbaum (1981, p. 349) report:

The network contains knowledge at all levels: acoustic, phonemic, lexical,

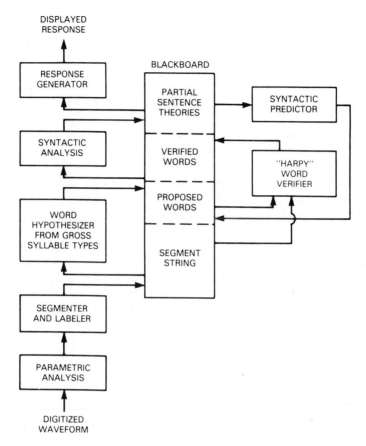

Figure 10-3 Block diagram of the CMU Hearsay II system organization. (From Klatt, 1977. Used by permission of the Acoustical Society of America.)

syntactic, and semantic. It stores acoustic representations of every possible pronunciation of the words in all of the sentences that HARPY recognizes. The alternative sentences are represented as paths through the network, and each node in the network is a template of allophones (distinctive variations of phonemes, dependent on adjacent phonemes).

The paths through the network can be thought of as "sentence templates," much like the word templates used in isolated-word recognition.

HARPY uses a heuristic method called *beam search* for searching for the sentence in the network that most closely matches the input signal. HARPY proceeds from left to right through the network, matching spoken sounds to allophonic states, and assigning scores based on the goodness of the match. HARPY keeps the paths with the best cumulative scores, pruning away others which fall some threshold amount below the best scoring path (Erman et al., 1980).

10-5.4. HWIM

The HWIM (Hear What I Mean) speech understanding system was developed at Bolt, Beranek and Newman. HWIM's domain was that of travel budget management. HWIM's organization is shown in Fig. 10-4. The lower components digitize the speech signal and generate a parametric representation of it, which is then segmented and labeled into phonemes which are ranked as to the quality of their match. These ranked phonemes are pictured as a segmented lattice, which is a graph that is divided into time segments and read from left to right. This graph is matched against a dictionary of word pronunciations (stored as a network with phonemes for nodes) by lexical retrieval components which generate word hypotheses.

HWIM's higher levels include information about trips (semantics), syntax, and word verification. The verification component takes the pronunciation of hypothesized words and generates a synthesized parameter representation that is compared to the parameters generated from the actual signal.

HWIM has a central control which uses the system's knowledge sources as subroutines. The system extends bottom-up theories using the top-down syntactic and semantic components. The system expands its hypotheses about the first recognized word in the sentence.

10-5.5. Summary of the ARPA SUR Program

ARPA's program did not result in a usable speech understanding system. The resulting systems were too slow, too restricted, and required large computational resources. However, it did discover and elucidate much new information about speech, and developed new architectural insights, particularly the blackboard architecture that has since been used in other AI systems. Performances of the different systems were difficult to compare because of the different vocabularies and domains employed. One critical factor in comparison is the *average branching factor* (ABF). This refers to the average number of words that might come next after each word in a legal sentence. Higher branching factors indicate a greater diversity of possible expressions. Table 10-3 summarizes the three major ARPA SUR projects. Note that the ABF is 196 for HWIM's data base retrieval task, versus 33 for Hearsay's and HARPY's document retrieval task.

10-6. STATE OF THE ART

10-6.1. Speech Recognition

Table 10-4 is a summary of a recent Texas Instruments' study of commercial speech recognizers tested on a 20-word vocabulary consisting of the 10 spoken digits

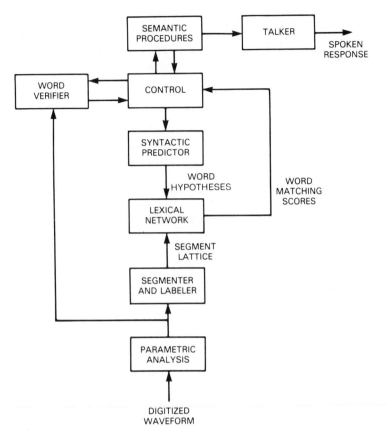

Figure 10-4   Block diagram of the BBN HWIM system organization. (From Klatt, 1977. Used by permission of the Acoustical Society of America.)

zero through nine and 10 command words: start, stop, yes, no, go, help, erase, rubout, repeat, and enter.

In 1982, speaker-dependent connected-word short-string, small vocabulary (approximately 50 words) recognizers were commercially available. These could recognize up to 90 words per minute (wpm) of connected speech compared to a typical person's speaking rate of 150 wpm. The vocabulary size is usually fewer than 150 words but is application dependent. Recognition accuracies of 98% or greater are being achieved in factory environments. Turnkey systems were in the range $5000 to $75,000 in 1983. Consumer product speech-recognizer subsystems for toys, personal computers, voice-controlled appliances, and so on, cost from $6 to $100.

Voice recognition systems are here, viable, proven, but still somewhat costly. In industrial applications, they have demonstrated large increases in productivity. Hundreds of successful installations exist today. Pluhar (1983) indicates the human factor considerations associated with successful applications.

TABLE 10-3 Summary of ARPA's Speech Understanding Systems

Name (Organization)	Domain/ Purpose	Approach	Knowledge Representation	Control	ABF	Accuracy (%)	Comments
Hearsay II (Carnegie-Mellon University)	AI publications document retrieval	Utilizes cooperating independent system experts (knowledge sources) that communicate via posting hypotheses on a blackboard	Independent KSs composed of production rules	Asynchronous pattern-invoked knowledge sources Opportunistic scheduling by first expanding the highest scoring hypothesis	33	90	Development of blackboard architecture and use of independent cooperating knowledge sources most significant A parallel processor version has been built to exploit KS modularity
HARPY Carnegie-Mellon University)	AI publications document retrieval	Compiled a network of all possible pronunciations of all possible sentences; paths through network are "sentence templates"	Precompiled network; each node is a template of allophones, which when linked form acoustic representations of every possible pronunciation of words in the domain	Beam search No backtracking	33	95	Approach cannot easily accommodate pragmatics Needs a large memory Sensitive to missing acoustical segments and missing words

HWIM (Bolt, Beranek and Newman)	Travel budget management	Extends bottom-up word theories, using top-down syntactic and semantic components; verifies hypothesized words by generating a parameter representation that is compared with that from the actual speech input. Uses an ATN semantic grammar	Uses networks to represent: Trip facts and relations Lexicon Phoneme hypotheses from signal	Centralized control using KSs as subroutines Expands sentences about the first recognized word in sentence (island driving)	196	44	Speaker independent Very slow Most difficult domain in SUR project

TABLE 10-4 T.I.'s Test of Speech Recognizers on Individual Words

Manufacturer	Model[a]	Nominal Price in 1981 (thousands)	Nominal Price for Comparable 1983 Model[b] (thousands)	Percent Substitutions
Verbex	1800	$65	$19.6	0.2
Nippon Electric	DP-100	$65	$27	1.2
Threshold Technology	T-500	$12	$ 5	1.4
Interstate Electronics	VRM	$ 2.4	$ 2.4	2.9
Heuristics	7000	$ 3.3	NA	5.9
Centigram	MIKE 4725	$ 3.5	NA	7.1
Scott Instruments	VET/1 (home computer peripheral)	$ 0.5	$ 0.5	12.6

[a] First two systems are capable of connected speech. Verbex is the only system having speaker-independent capability.
[b] NA, not applicable.

Source: After Doddington and Schalk (1981).

10-6.2. Speech Understanding

In 1983 there were no commercial true speech-understanding systems. However, there are a number of U.S. companies working on such systems.

Bell Labs.   Bell has been working on a semantic sentence recognizer and interpreter utilizing a finite-state grammar and a small vocabulary. The intent is to produce an interactive speech understanding system for use over the telephone (Levinson and Liberman, 1981).

IBM—T. J. Watson Research Center.   IBM has had one of the largest long-standing efforts in continuous-speech recognition and understanding.

Other organizations involved in developing speech understanding systems include Bolt, Beranek and Newman.

10-7. PLAYERS AND RESEARCH TRENDS

10-7.1. Who Is Doing Speech Recognition Work

- Commercial organizations
    IBM
    Texas Instruments
    Bell Labs.
    Verbex
    Nippon Electric
    Threshold Technology
    Interstate Electronics
    Matsushita
    Scott Instruments
    Sanyo
    Intel
    ITT (San Diego)
    Fairchild
    Hewlett-Packard
    Haskins Lab.
    Lincoln Labs.
    Speech Communications Research Lab.
    Sperry Univac
    Votan
    Voice Machine Communications
    Voice Processing Corp.
    General Instrument—Milton Bradley
    Voice Control Systems

- Universities
    MIT
    Carnegie-Mellon University
    Virginia Polytechnic Institute
    University of California, Berkeley

10-7.2. Research Trends

Research is proceeding along the lines of improved methods for determining end points of words, removing noise, separating linguistically significant variations in the speech signal from insignificant variations (such as variations in word pronunciations), increasing processing speed, improving system architectures, and developing specialized hardware.

10-8. FUTURE DIRECTIONS

It is anticipated that speaker-independent, continuous-speech recognition systems with limited vocabularies (10 to 20 words), having an accuracy of 98% or better, will be available in the mid-1980s. Automatic dictation will probably not appear before the 1990s, with Japanese-language systems being the first to appear. (The Japanese language has only about 500 syllables, compared to 10,000 for English). Speech understanding is a major part of the Japanese Fifth-Generation Computer project (Feigenbaum and McCorduck, 1983).

Due to the advancement in VLSI, it is expected that expanded voice recognition chips for toys and similar products will soon be in the $5 range—less than $50 for a complete system.

A strong expectation is that a speech-understanding system using a natural language parser will be introduced by IBM in the late 1980s. Around 1990, true commercial speech-understanding systems, having the capabilities of the ARPA SUR systems but operating in near real time, are expected to appear. By 1990, speech recognition and understanding is expected to be a billion-dollar-a-year industry (Elphick, 1982).

REFERENCES

Arden, B. W., (Ed.), *What Can Be Automated?* Cambridge, MA: MIT Press, 1980.

Barr, A., and Feigenbaum, E. A., (Eds.), *The Handbook of Artificial Intelligence*, Vols. 1 and 2. Los Altos, CA: W. Kaufmann, 1981, 1982.

Doddington, G. R., and Schalk, T. B., "Speech Recognition: Turning Theory to Practice," *IEEE Spectrum*, Sept. 1981, pp. 26–32.

Elphick, M., "Unraveling the Mysteries of Speech Recognition," *High Technology*, Vol. 2, No. 2, Mar./Apr. 1982, pp. 71–76.

Erman, L. D., Hayes-Roth, F., Lesser, V. R., and Reddy, D. R., "The Hearsay-II Speech–Understanding System: Integrating Knowledge to Resolve Uncertainty," *Computing Surveys*, Vol. 12, No. 2, 1980.

Feigenbaum, E. A., and McCorduck, P., *The Fifth Generation*. Reading, MA: Addison-Wesley, 1983.

Klatt, D. H., "Review of the ARPA Speech Understanding Project," *Journal of the Acoustical Society of America*, Vol. 62, No. 6, Dec. 1977, pp. 1345–1366.

Lea, W. A., and Shoup, J. E., *Review of the ARPA SUR Project and Survey of Current Technology in Speech Understanding*, Speech Communications Research Lab., Los Angeles, Jan. 16, 1979.

Levinson, S. E., and Liberman, M. Y., "Speech Recognition by Computer," *Scientific American*, Vol. 244, No. 4, Apr. 1981, pp. 64–76.

Pluhar, K., "Speech Recognition–An Exploding Future for the Man–Machine Interface," *Control Engineering*, Jan. 1983, pp. 70–73.

Zue, V. W. (Haton, J-P. editor), "Acoustic–Phonetic Knowledge Representation: Implications from Spectrogram Reading Experiments," *Proceedings of the 1981 NATO Advanced Summer Institute on Automatic Speech Analysis and Recognition*. Boston: D. Reidel, 1981.

Zue, V. W., "Tutorial on Natural Language Interfaces: Part 2–Speech," *AAAI-82*, Aug. 17, 1982, Pittsburgh, Pa.

11

APPLICATIONS OF AI

The potential range of AI applications is so vast that it covers virtually the entire breadth of human intelligent activity. Detailed listings of focused applications are given in each of the chapters on applications: expert systems, computer vision, natural language processing, planning, and speech recognition and speech understanding. This chapter just summarizes some of the key applications. Generic applications are listed in Table 11-1. Examples of specific applications of AI are listed in Table 11-2. Potential functional applications for NASA (the National Aeronautics and Space Administration) are indicated in Table 11-3. The opportunities this opens up for NASA are listed in Table 11-4. Similar opportunities are available in many other public and private domains.

TABLE 11-1   Generic Applications of AI

Knowledge management
  Intelligent data base access
  Knowledge acquisition
  Text understanding
  Text generation
  Machine translation
  Explanation
  Logical operations on data bases
Human interaction
  Speech understanding
  Speech generation

Learning and teaching
  Intelligent computer-aided instruction
  Learning from experience
  Concept generation
Fault diagnosis and repair
  Human beings
  Machines
  Systems
Computation
  Symbolic mathematics
  "Fuzzy" operations
  Automatic programming
Communication
  Public access to large data bases via telephone and speech understanding
  Natural language interfaces to computer programs
Operation of machines and complex systems
Autonomous intelligent systems
Management
  Planning
  Scheduling
  Monitoring
Sensor interpretation and integration
  Developing meaning from sensor data
  Sensor fusion (integrating multiple sensor inputs to develop high-level interpretations)
Design
  Systems
  Equipment
  Intelligent design aids
  Inventing
Visual perception and guidance
  Inspection
  Identification
  Verification
  Guidance
  Screening
  Monitoring
Intelligent assistants
  Medical diagnosis, maintenance aids, and other interactive expert systems
  Expert system building tools

TABLE 11-2   Examples of Domain-Specific Applications of AI

Medical
  Diagnosis and treatment
  Patient monitoring
  Prosthetics
    Artificial sight and hearing
    Reading machines for the blind
  Medical knowledge automation
Science and engineering
  Discovering
    Physical and mathematical laws
    Determination of regularities and aspects of interest
  Chemical and biological synthesis planning
  Test management
  Data interpretation
  Intelligent design aids
Industrial
  Factory management
  Production planning and scheduling
  Intelligent robots
  Process planning
  Intelligent machines
  Computer-aided inspection
Military
  Expert advisors
  Sensor synthesis and interpretation
  Battle and threat assessment
  Automatic photo interpretation
  Tactical planning
  Military surveillance
  Weapon-target assignment
  Autonomous vehicles
  Intelligent robots
  Diagnosis and maintenance aids
  Target location and tracking
  Map development aids
  Intelligent interactions with knowledge bases
International
  Aids to understanding and interpretation
    Goals, aspirations, and motives of different countries and cultures
    Cultural models for interpreting how others perceive
  Natural language translation
Services
  Intelligent knowledge base access: airline reservations
  Air traffic control
  Ground traffic control
Financial
  Tax preparation
  Financial expert systems
  Intelligent consultants

Executive assistance
  Read mail and spot items of importance
  Planning aids
Natural Resources
  Prospecting aids
  Resource operations
    Drilling procedures
    Resource recovery guidance
  Resource management using remote-sensing data
Space
  Ground operations aids
  Planning and scheduling aids
  Diagnostic and reconfiguration aids
  Remote operations of spacecraft and space vehicles
  Test monitors
  Real-time replanning as required by failures, changed conditions, or new opportunities
  Automatic subsystem operations

TABLE 11-3   Potential Functional Applications of AI in NASA

Planning and scheduling

Test and checkout

Symbolic computation

Information extraction

Operations management
  Monitoring
  Control
  Sequencing

System autonomy
  Subsystem management
  Fault diagnosis

Intelligent assistants

Space operations requiring autonomy due to time delays that prohibit teleoperation.

TABLE 11-4   AI and NASA

AI opens up an opportunity for NASA to:

  Dramatically:
    Reduce costs
    Increase productivity
    Improve quality
    Raise reliability
    Utilize facilities and people more effectively
  Provide new mission capabilities
  Enable new missions
  Improve aerospace science and technology

by using AI techniques to increase human productivity and to help automate many activities
previously requiring human intelligence.

12

WHAT DOES IT ALL MEAN?

12-1. AN OVERVIEW OF THE STATE OF THE ART IN AI

12-1.1. General

The state of the art of AI is moving rapidly as new companies enter the field, new applications are devised, and existing techniques are formalized. The cutting edge of AI today is expert systems, with hundreds of prototype systems having been built. With the advent of personal LISP machines and the general reduction in computing costs, development of commercial AI systems are now under way. A number of natural language interfaces and computer vision systems are already on the market.

Japan has focused on AI capabilities as the basis for its Fifth-Generation Computer (Warren, 1982), and has already initiated research toward this $500 million, 10-year goal. In response, Britain's Department of Industry has formed a study group to coordinate British AI efforts, and the ALVEY Programme for Advanced Information Technology has been established. The European Common Market Community has established the European Strategic Programme on Research in Information Technology (ESPRIT). In the United States, DARPA, which had been spending about $20 million annually on AI research, is now dramatically expanding its efforts in relation to fifth-generation computers with the establishment of the SCS (Strategic Computers and Survivability) superintelligent computer project. The U.S. Navy, Army, and Air Force are all initiating substantial AI efforts. The Navy has established a major tri-service AI applied research center at Bolling AFB. The Air Force is focusing its in-house AI research efforts at Rome Air Development Center and Wright-Patterson AFB. The Army has established the University of Pennsylvania and the University of Texas at Austin as AI centers of excellence in response to its perceived needs.

Over a dozen U.S. computer companies have also responded to the Japanese challenge by forming MCC (the Microelectronics and Computer Technology Corp.) in Austin, Texas.

12-1.2. Basic Core Topics

AI basic theory and techniques are now being codified. The *Handbook of Artificial Intelligence* (Barr and Feigenbaum, 1981, 1982; Cohen and Feigenbaum, 1982) (funded by DARPA) has been a major contribution in pulling together the basic theory and making it available at the graduate level for students and practitioners of AI.

Search theory is now relatively mature and well documented. A number of knowledge representation techniques have been devised and are now supported by representation languages. Basic programming languages have continued to evolve, with INTERLISP being the best supported, but Common LISP is beginning to emerge from MIT's MACLISP as the language of AI personal computers. PROLOG, a logic-based programming language, popular in Europe, appears to be the language of choice for Japan's Fifth-Generation Computer and is beginning to awaken interest in the United States. PROLOG holds promise of reinvigorating first-order predicate logic as a major factor in AI for knowledge representation and problem solving.* A large number of problem-solving techniques developed during the last two decades are now forming the basis for the inference engines in expert systems. Although much work remains to be done, the core topics of AI are now in a sufficient state of readiness for use in initial AI applications.

12-1.3. Expert Systems

Although expert systems are no longer a rarity, at the start of 1984 only a few, such as MOLGEN, R1, ONCOCIN, DENDRAL, Dip-Meter Advisor, ACE, and PUFF, were in actual commercial use on a regular basis. Expert systems are still restricted to a narrow domain of expertise and require laborious construction via interaction with human experts. Further, these systems tend to have the characteristics of:

- Providing satisfactory rather than optimum solutions
- Providing satisfactory answers only some of the time (as little as 60% in some medical diagnosis systems, although on the order of 99% for R1/XCON)
- Being unable sometimes to provide an answer

This is in contrast to normal engineering solutions that are algorithmic in nature

---

*Additional information on LISP and PROLOG, together with illustrative examples, is given in Appendix C.

and virtually always provide a satisfactory answer when supplied with appropriate inputs. This "sometimes the answer is wrong" characteristic of expert systems is also characteristic of human decision making. Thus, at the moment, expert systems tend to be used as human assistants (with human beings making the final decisions) rather than as "stand-alone" autonomous systems.

12-1.4. Natural Language

Natural language interfaces (NLIs) were the first commercial AI product. ROBOT (now INTELLECT, by the Artificial Intelligence Corp., Waltham, Massachusetts) in 1980 was the first on the market and new exists in over 200 installations. Some half-dozen other commercial NLI systems are now available. All these systems are restricted to limited sets of natural language and exhibit occasional failures in understanding or processing a user's input. However, with a little training of the users, NLIs have proved very useful.

Several commercial machine translation systems are also available. These are not used as completely autonomous systems, as in many cases their translation is very rough or even incorrect. (The classic example is "The spirit is willing but the flesh is weak," translated into Russian as "The vodka is good but the meat is spoiled.") However, as an aid to a human translator, they can improve productivity by a factor of 2 to 10, depending on the system and the material being translated.

Text understanding and text generation are still in the research stage.

12-1.5. Computer Vision

Computer vision has entered the commercial market, with dozens of companies offering commercial vision systems. These systems are operating successfully in specialized environments on low-level problems of verification, inspection, recognition, and determination of object location and/or orientation. Current commercial vision systems deal primarily with two-dimensional images—they cannot handle three-dimensional analysis, needed to recognize objects from arbitrary viewpoints.

Although quite a number of high-level research vision systems have been explored, no general vision system is available today or is imminent. Major current efforts in this area are ACRONYM at Stanford University, VISIONS at the University of Massachusetts, and the robotic vision effort at the National Bureau of Standards.

12-1.6. Conclusions

Figure 12-1 is a list of overall conclusions on the current state of AI. Summarizing, it appears that technology is now ready for early applications. However, the fact that current AI systems are prone to error suggests that current AI applications should be focused on intelligent aids for human beings rather than on truly autonomous systems.

- Basic principles and techniques devised and demonstrated

- Initial languages, programs and tools developed

- Software portability a problem

- A few initial applications already in use

- Technology is now ready for early applications

- Current technology more appropriate for intelligent assistants than for autonomous systems

- Customizing, adapting and usually writing own programs necessary

- Because of huge potential benefits, utilization will be explosive as technology is further rationalized

Figure 12-1 Conclusions on the current state of the art in AI.

12-2. WHAT IS LACKING

Deficits in AI are:

1. *Sufficient people adequately trained in AI:* The rise of AI companies has depleted the teaching and research at the principal AI centers at universities and research institutes.
2. *Good system building tools:* Commercial expert systems building tools are just beginning to emerge. Examples are LOOPS from Xerox PARC, KEE from Intellicorp and S.1 from Teknowledge in Palo Alto, California, ARBY from Smart Systems Technology in McLean, Virginia, Expert-Ease from Intelligent Terminals Ltd. in Great Britain, and ART (Advanced Reasoning Tool) from Inference Corp. in Los Angeles.
3. *Transportable software:* Most expert systems are still not in good software form, fully documented, and portable.
4. *Good user interfaces to AI software systems:* As yet, few expert systems or other AI systems have user-friendly interfaces, most using very limited or no natural language interfaces.
5. *Error-free performance:* Expert systems and natural language interfaces are still error-prone, more appropriate for intelligent assistants than for truly autonomous systems.
6. *Good explanation capabilities:* Most expert systems have as yet either none or very stilted explanation systems, such as simply listing the rules used in arriving at the solution.
7. *Understanding of AI by managers:* Thus far, very few engineering or research managers understand AI.
8. *Systems that learn from experience:* As yet virtually none of the operational

expert systems exhibit the learning needed to improve their performance or to develop their knowledge base, though this is beginning to change.

12-3. WHAT SHOULD BE DONE

- There is a strong need for supporting more teaching and research in AI at universities. This can be aided by government programs and grants from large corporations. It can also be facilitated by encouraging AI personnel in industry to teach part-time.
- Good system tools are beginning to evolve, but again encouragement by the government might be helpful.
- Transportability of software. This will improve as a common AI programming environment such as Common LISP (being promoted by DARPA) becomes more of a reality.
- Natural language interfaces (NLIs). The coming availability of transportable NLIs will be an aid in achieving good explanation facilities and user-friendly interfaces.
- Improved performance of AI systems. The still errorful performance of AI systems will be alleviated as increased causal knowledge is added, relying less exclusively on compiled experience.
- Increased effort in achieving self-learning systems. This can probably best be supported through further government encouragement of learning research at universities and research institutes.

12-4. THE FUTURE

12-4.1. General

Today's initial AI systems can be regarded primarily as intelligent assistants. These are taking the form of expert systems, natural language interfaces, computer vision systems, and intelligent computer-aided instruction systems. They—like human beings—are prone to failure, but unlike human beings, they are not capable of drawing on deep knowledge when needed to achieve graceful degradation, so that their failures are more abrupt. Thus researchers are currently engaged in developing a new set of advanced systems, based on deep knowledge—which includes such aspects as causal models and scientific knowledge.

12-4.2. Expert Systems

Utilizing emerging expert-system building tools, AI developers are expected eventually to put expert medical, financial, and legal advice at the fingertips of anyone with access to a personal computer (although this will probably have to await the arrival of a new generation using 32-bit microprocessors). Expert systems will also put expertise in the hands of less-trained, lower-salaried workers.

12-4.3. Natural Language

Speech recognition appears to be emerging as a key human-machine interface. Researchers have found that the psychological problems inherent in talking to a machine are a barrier to the acceptance of speech interfaces. Overcoming the psychological problems may thus be an important factor. To achieve practical continuous speech recognition, systems will have to expand today's vocabularies by an order of magnitude, increase speed by two orders of magnitude, and get costs below $1000. Such systems are estimated to still be at least five years away.

Natural language interfaces appear to be the way to vastly increase the number of people who can interact with computers. Systems with near natural language capabilities are available now, although it will be years before the systems can handle truly unrestricted dialogue. It is estimated that public access to large data bases via computer using restricted speech understanding will begin to appear early in the second half of this decade.

This can be expected to open up a whole new industry of automated reservation, shopping, and information services assessed by telephone.

Another emerging aspect of natural language processing are systems that understand text by utilizing world knowledge. Such systems could read and summarize news stories (as is now being done in research) but more likely would be applied to such tasks as reading mail and informing the recipient of important items, or in general, processing large amounts of information for human beings trying to escape from overload.

12-4.4. Computer Vision

Computer vision will increasingly be used in industry for inspection, identification, verification, part location, and other purposes. Vision provides the most general-purpose sensory input for intelligent robots. It is likely that roughly 25% of all robots will utilize vision by the end of the decade. Vision is also expected to play a large part in military automation, remote sensing, and as aids to the handicapped.

12-4.5. Intelligent Robots

The development of AI is making intelligent robots feasible. As intelligence is added to robots, they will not only be able to perform more flexibly in manufacturing, but will begin to be evident in tasks outside the industrial environment. Thus robots in firefighting, underseas exploration, mining, and construction will appear. However, the big push may be in military applications with its actively hostile environments. In the 1990s, robots with intelligence and sensory capabilities will appear in the service industries—in everything from food service to household robots. It is also anticipated that in the 1990s, intelligent robots will enter the space arena for such tasks as the construction and assembly of large space structures; space manufacture; extraterrestrial mining and exploration; and operation, maintenance, and repair of space installations.

12-4.6. Industrial Applications

In addition to more intelligent robots, AI will influence virtually every aspect of the future industrial plant. Integrated plants that make use of automated planning, scheduling, process control, warehousing, and the operation of automated robot carts, robots, and manufacturing machines will appear in a few years and will become widespread within the next 10 years.

12-4.7. Computers for Future Automation

Computers and special-purpose chips designed to incorporate parallel processing are being developed at several universities and computer organizations. MIT has been developing a parallel machine using VLSI techniques to break problems into subproblems and distribute them among its processors. Another chip will utilize parallel processing to search rapidly through the branches of a semantic network.

The most prominent future system is Japan's Fifth-Generation Computer that could store and retrieve some 20,000 rules, incorporate a knowledge base of 100 million data items, and help make Japan an AI leader before the end of this century. To help maintain U.S. competitiveness, more than a dozen of the largest U.S. electronics and computer companies have recently set up the Microelectronics and Computer Technology Corp. (MCC) in Austin, Texas. This well-funded cooperative research venture is designed to develop a broad base of fundamental technologies. Among them is a 10-year program to develop advanced computer architectures and artificial intelligence. Similar responses have occurred in Great Britain and Europe.

12-4.8. Computer-Aided Instruction

Intelligent CAI systems may produce one of the most dramatic changes of all. Education consumes some 10% of the U.S. gross national product today. Systems that will enable students to ask questions and receive insightful answers may begin to overcome the barriers of instruction by machines. Computer systems that model the student based on his or her response can gear instruction to the student's level of ability and interest, something not easily done in a conventional classroom.

To truly learn is to digest and make the material one's own by updating one's internal models and using them in new applications. Real-time interaction with a computer providing immediate feedback and individual guidance is particularly appropriate to this goal.

Thus, as computer hardware costs continue to tumble, the nature of the entire present educational system may be radically changed. For adults and members of the armed forces, CAI will probably rapidly become the standard form of instruction.

12-4.9. Learning by Computers

The real breakthrough may come when machine learning is achieved. Already, several learning systems, currently in the research stage, have been able to produce very interesting results (Carbonell, et al., 1983). Someday machines will be able to learn throughout their lifetime, building up the knowledge base needed for advanced reasoning. This will open up spectacular new applications in offices, factories, and homes.

Machines may update their knowledge by reading natural language material, as well as learning by experience from the problems the computers are called upon to solve. Computers may also be able to form conclusions from an examination of multiple data bases, thereby building new knowledge from existing knowledge.

12-4.10. The Social Impacts

The U.S. Defense Science Board has ranked AI as one of the top 10 military technologies of the 1980s. Not only will human-level expertise and decision-making capabilities show up in machines, but the task of achieving these results will help us to understand how our minds work as well.

Combining expert systems and computer graphics will enable people to "see" the results of the computer actions. This will not only clarify and simplify the interaction, but will greatly speed human learning and decision making. The result may be to compress months of research and engineering experience gained the old way into insights gathered from just a few hours of interaction with intelligent computer programs.

AI's effects on society may be slow at first, but by the end of the century the results should be revolutionary. The shift in employment away from manufacturing may be as dramatic as the shift away from agriculture. There will also be a revolution in white-collar work—service, research, leisure. How to restructure society to take advantage of a potential abundance of goods and services, or to adapt to new work opportunities and leisure activities, may be the question of the century. This may give society another chance to pursue the social and mental goals so often deferred. It may also free us at last from the monetary and technical bonds to earth. Perhaps, we can at last "reach for the stars."

REFERENCES

Barr, A., and Feigenbaum, E. A. (Eds.), *The Handbook of Artificial Intelligence*, Vols. 1 and 2. Los Altos, CA: W. Kaufmann, 1981, 1982.

Carbonell, J. G., Michalski, R. S., and Mitchell, T. M., "Machine Learning: A Historical and Methodological Analysis," *AI Magazine*, Vol. 4, No. 3, Fall 1983, pp. 69–79.

Cohen, P. R., and Feigenbaum, E. A. (Eds.), *The Handbook of Artificial Intelligence*, Vol. 3. Los Altos, CA: W. Kaufmann, 1982.

Warren, D. H. D., "A View of the Fifth Generation Computer and Its Impact," *AI Magazine*, Vol. 3, No. 4, Fall 1982, pp. 34–39.

Part II

ROBOTICS

13

WHAT CAN WE CALL A ROBOT?

Robots in use today are primarily machines with manipulators that can easily be programmed to do a variety of manual tasks automatically.* The robot consists of the following:

- One or more manipulators (arms)
- End effectors (hands)
- A controller
- And, increasingly, sensors to provide information about the environment and feedback of performance of task accomplishment

The motivation for most current robotic development is industrial, to achieve the following goals:

- Increase productivity
- Reduce costs
- Overcome skilled labor shortages
- Provide flexibility in batch manufacturing operation
- Improve product quality
- Free human beings from boring and repetitive tasks, or operations in hostile environments.

*The Robot Institute of America (RIA, now the Robotic Industries Association) definition is: "A robot is a reprogrammable multi-functional manipulator designed to move material, parts, tools, or specialized devices, through variable programmed motions for the performance of a variety of tasks."

Table 13-1 gives the Japanese standard classifications for industrial robots (IRs) by input information and teaching method, and relates this to the U.S. classification by control method: nonservo or servo. It can be observed that the Japanese view of robots is less restricted than the U.S. view.*

*All references for Part II will be found following Chapter 28.

TABLE 13-1   Definitions and Classifications of Industrial Robots

Japanese Industrial Standard JIS B0134-1979	U.S. View
Manipulator: A device for handling objects as desired without touching with the hands, and it has more than two of the motional capabilities, such as revolution, out-in, up-down, right-left traveling, swinging or bending, so that it can spatially transport an object by holding, adhering to, and so on. *Robot:* a mechanical system which has flexible motion functions analogous to the motion functions of living organisms or combines such motion functions with intelligent functions, and which acts in response to the human will. In this context, intelligent functions mean the ability to perform at least one of the following: judgment, recognition, adaptation, or learning.	*Industrial robots* are primarily machines with manipulators that can be easily programmed to do a variety of manual tasks automatically. In Japan, also included under industrial robots are manual manipulators operated by human beings directly or indirectly. In the U.S., the latter are usually referred to as teleoperators.

Classification by input information and teaching method

Classification by servo type.

Manual manipulator: a manipulator that is directly operated by a human being. *Sequence robot:* a manipulator, the working step of which operates sequentially in compliance with preset procedures, conditions, and positions. *Fixed sequence:* a sequence robot as defined above, for which the preset information cannot easily be changed. *Variable sequence:* a sequence robot as defined above, for which the preset information can be easily changed. *Playback robot:* a manipulator that can repeat any operation after being instructed by a human being. *Numerically controlled robot:* a manipulator that can execute the commanded operation in compliance with the numerically loaded working information on e.g., position, sequence, and conditions. *Intelligent robot:* a robot that can determine its own actions through its sensing and recognitive abilities.	*Nonservo robots.* Under this classification are open-loop robots which do not incorporate feedback to control the manipulator's path. The simplest example are *pick and place robots* with only two or three degrees of freedom used to transfer items from place to place by means of point-to-point moves. *Servo robots,* which include all the categories described below, are the most common U.S. robots. These robots are controlled by servomechanisms which sense the difference between the commanded trajectory and the actual trajectory and act to reduce this difference. Four to seven mechanical degrees of freedom are commonly used to enable the arm end-effector to achieve the range of positions and orientations desired. *Record-playback robot* is a servo robot directed by a programmable controller. There are two basic types of control: 1. *Point-to-point control.* In this approach, the controller first records the manipulator joint positions corresponding to the sequence of critical points along the desired task trajectory as the robot is moved under operator control. To perform the task, the robot servo system uses these as guide points and generates the intervening segments to achieve the final trajectory. 2. *Continuous path control.* In this control scheme, the inputs are specified at every point along the desired path of motion. This is usually achieved by the operator physically leading the robot through the desired actions as the recording takes place. *Computer-controlled robot.* This is a servo robot directed by a computer, thus providing the capability for great flexibility. Though this type of robot can also be programmed by leading the robot through the desired actions, the tendency today is toward off-line programming using higher order languages. *Sensory-controlled robot.* A robot whose program sequence can be modified as a function of information sensed from its environment. Though such a robot can be servoed or non-servoed, the current trend is toward computer control with the robot making performance choices contingent on sensory (e.g., visual or tactile) inputs. *Assembly robot.* This is a recent category of robots, specifically designed for assembly-line applications. These robots are computer-controlled, usually incorporating sensory feedback.

Sources: Japanese standard: Yonemoto, 1981, p. 5. Used by permission of the Japan Industrial Robot Association (JIRA), Tokyo. U.S. view: Based on RIA Robotics Glossary, 1984.

14

WHAT CAN ROBOTS DO?

As indicated in Table 14-1, the Japanese are the world's largest user of robots. Of the 1982 estimate of 55,000 robots in operation in the entire world, Japan had well over half, while the United States had less than 12%. Table 14-2 presents the Japanese robotic use by industry, based on robot costs. It appears that the electrical machinery and automobile industries are their largest users of robots. Less than 1% of IRs were utilized outside of manufacturing. Table 14-3 compares Japanese and U.S.

TABLE 14-1 IRs in Operation (Approximate
Number as of the End of 1982)[a]

1. Japan	31,900
2. United States	6,300
3. West Germany	4,300
4. Sweden	1,500
5. Italy	1,100
6. France	1,000
7. United Kingdom	1,000
8. Belgium	300
9. Canada	300
10. Poland	300

[a] Manual manipulators and fixed-sequence robots are not included.

Source: Worldwide Robotics Survey and Directory (1983). Used by permission of the Robot Institute of America.

TABLE 14-2   **Principal Users of Industrial Robots in Japan**
(Japanese Definition of Robot)

	Share in 1982 (Based on Value) (%)
Electric/electronic industry	30
Auto industry	27
Machinery industry	7
Metal products industry	5
Other	17
Exports	14

Source: Japanese Industrial Robot Association (Aron, 1983, p. 42). Used by permission of Paul H. Aron, Vice Chairman of the Board of Directors, Daiwa Securities America Inc., New York, N.Y., and Professor of International Business at the Graduate School of Business Administration of New York University, NYC.

TABLE 14-3   **Percentage of Robot Usage by Production Process**
(End of 1982)

Process	Japan	U.S.
1. Welding	25	38
2. Material handling	21	21
3. Machine loading and unloading	8	17
4. Assembly	19	1
5. Casting	2	14
6. Painting and finishing	4	8
7. Other	21	1
	100	100

Source: Worldwide Robotics Survey and Directory (1983). Used by permission of the Robot Institute of America.

robot usage by production process. It is apparent that welding and moving or positioning a workpiece dominates robot use. It also appears that Japan is preceding the United States in substantial use of robots for assembly. This is reflected in the production of Japanese robots by type, where we note from Table 14-4 that by 1982, intelligent robots had already reached 8% of production, up from only 1% two years earlier.

Of the many reasons for installing industrial robots, the most prominent ones are to increase productivity and reduce costs. A commonly used measure for determining the economics of installing robots is the payback period. This is usually given (see, e.g., Simons, 1980, p. 32) as the cost of the robot and accessories (including fixtures, etc.) divided by the annual savings:

TABLE 14-4   Japanese Production of Robots by Type (1982)

Type	Units	
Manual manipulators	1,200	(5%)
Fixed-sequence control robots	8,600	(35%)
Variable-sequence control robots	4,300	(17%)
Playback robots	6,700	(27%)
Numerically controlled robots	2,000	(8%)
Intelligent robots	2,000	(8%)
	24,800	(100%)

Source: Aron (1983, p. 37). Used by permission of Paul H. Aron, Vice Chairman of the Board of Directors, Daiwa Securities, America Inc., New York, N.Y. and Professor of International Business at the Graduate School of Business Administration of New York University, NYC.

$$p = \frac{I}{L - E}$$

where p = payback period, years

I = total investment

L = total annual savings in labor

E = total annual robot upkeep cost

Typically, automation investments have required that the payback period be three years or less, with actual payback periods reaching as low as one year when multiple shifts are utilized.

15

PRIMITIVE FUNCTIONS ACCOMPLISHED

BY A ROBOT ARM

Industrial robots are designed primarily for manipulation purposes. The primitive actions performed by the manipulator are:

1. Moving
   a. From point to point
   b. Following a desired trajectory
   c. Following a contoured surface
2. Grasping
   a. Parts
   b. Tools
3. Sensing
   a. Touch
   b. Force

Nonservo robots can provide a sequence of point-to-point (PTP) motions. For each motion, the manipulator members move full tilt until the limits of travel are reached. Thus they are often referred to as *bang-bang, pick-and-place*, or *limited-sequence robots*. "Programming is done by setting up the desired sequence of moves and by adjusting the end stops for each axis" (Tanner, 1981, p. 7). The use of mechanical stops and limit switches yields good positional accuracy, which is typically repeatable to better than ±0.5 mm (Simons, 1980, p. 20).

Servo-controlled robots can be controlled to stop at arbitrary intermediate points, as well as move with controlled velocities. They can also be used to follow a programmed trajectory, or by using sensors to follow a contoured surface.

If we consider end effectors, the robot can also be made to grasp, push and pull, twist, use tools, perform insertions and assembly, and do other manipulations done by human beings. If sensors and appropriate computations are added, the robot can perceive, inspect, recognize, test, and do many of the other perceptual functions now done by humans.

All applications require that the robot interact with its environment in the execution of its programmed task. Thus the robot is often designed to switch machines on and off and to accept information from machines as to their status. The robot may also have provisions for accepting signals from a moving assembly line, so it can track it. Similarly, the robot may have interface provisions for tactile, visual, or other sensory inputs for feedback control.

16

ROBOT ARMS

16-1. MANIPULATOR ARM CONFIGURATIONS

To accomplish these primitive actions, four basic configurations of manipulator arm designs have evolved:

1. Rectangular [Fig. 16-1(a)]
2. Cylindrical [Fig. 16-1(b)]
3. Spherical [Fig. 16-1(c)]
4. Anthropomorphic—articulated or jointed arm [Fig. 16-1(d)]

These arms utilize both rotary ("revolute") and linear ("prismatic") joints, as appropriate, to accomplish their function. Six degrees of freedom (DOF) are necessary to locate the end effector (hand) at any point and any orientation in the work volume (6-DOF operation requires six joints). More DOFs are sometimes used to prevent interference of the arm with the workpiece or other obstacles. As few as 2 or 3 DOFs may suffice for a simple pick-and-place robot.

As indicated in Fig. 16-1 only 3 DOFs are required to position the end effector. However, for applications requiring a high degree of wrist flexibility, as many as three of the arms DOFs are located in the wrist, as shown in Fig. 16-2. These DOFs are usually referred to (using aeronautical terms) as *pitch, yaw*, and *roll*.

The basic four manipulator arm configurations can be combined in various ways to produce a large set of manipulator combinations. These combinations can be further extended by employing angled arms and offsets. Often, built-in mechanical

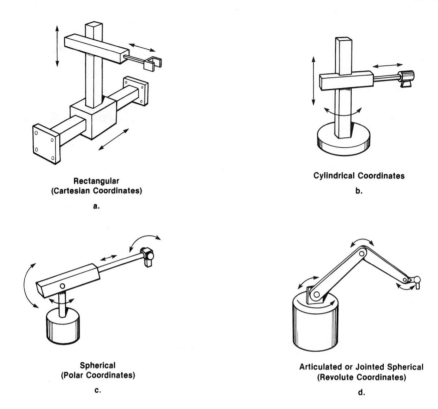

**Rectangular
(Cartesian Coordinates)**

a.

Cylindrical Coordinates

b.

**Spherical
(Polar Coordinates)**

c.

**Articulated or Jointed Spherical
(Revolute Coordinates)**

d.

Figure 16-1 Basic manipulator geometries

linkages are added to do such things as constrain the motion of nominally articulated
arms to cylindrical coordinates for ease of control.

   In addition, the arms may be designed to be mounted on their sides, hung
from the ceiling, or mounted on rails for mobility, to further multiply the range of
installed configurations.

16-2. ACTUATORS

There are three major ways today of powering the actuators that move the manipulator joints: pneumatic, hydraulic, and electrical. Pneumatic control is simple and
inexpensive but due to the compressibility (and therefore the impreciseness) of the
working medium, it is best used for pick and place robots whose trajectory motions
are controlled by mechanical stops. Hydraulic actuation is capable of high power
for a given size, but suffers from leakage problems as well as requiring pumps,

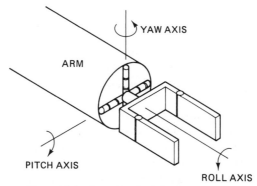

Figure 16-2   Example of a 3-DOF robot wrist.

storage tanks, and other accompanying paraphenalia. Electrical-mechanical actuation is clean but does not offer good power-to-weight ratios. Stepper motors, although relatively inexpensive, sometimes lose pulses and therefore accuracy. Direct-current (dc) torque motors are very reliable and offer good controllability but can be expensive. Because the overall cost advantage of hydraulic systems diminishes with size and power, electrical drives are preferable for the small robot. In terms of reliability, actuators are the most problem-prone area of robotics. Thus considerable research is needed in this area to improve reliability further [robots have an approximately 98% up-time (Engelberger, 1980, p. 117)] and to reduce costs and weight.

16-3. ROBOT HANDS (END EFFECTORS)

The *hand* or *gripper*, sometimes called the *end effector* (attached to the robot wrist) can be a mechanical, vacuum, or magnetic device for part (or object) handling or for tool manipulation. At present, most of these interchangeable devices are grippers, which are simple open-and-close devices. The variety of tools and grippers that can be adapted for robot use is virtually unlimited (see, e.g., Engelberger, 1980, pp. 41-58). The end effectors are usually unique to the robot application and thus customarily provided by the user,* although several manufacturers offer a broad selection of devices for grasping parts that might be adaptable to the particular task being performed. Nevertheless, although hundreds of different special-purpose end effectors now exist, the end effector remains one of the major limiting factors in universal robot manipulation due to lack of dexterity and programmability of the hands. Extensive research and development is now under way at institutions such as MIT and Stanford to produce grippers that can handle a wide variety of part configurations and to provide increased manipulative dexterity.

*This customization has sometimes proved to be very costly.

16-4. CONTROL

Control of robot arms ranges from highly repeatable open-loop devices to servo-controllers that utilize external sensors to control robot actions. Open-loop devices may be as simple as a step sequencer and positionable mechnical stops for a pick-and-place robot, to more complex devices using stepper motors or timing to reproduce a desired motion.

Servo devices can use internal sensors, such as joint-position sensors, or external information sensors such as force, proximity devices, and even vision (operating under computer control). Servos operating only on internal sensors require very careful positioning of the workpiece (which can require expensive fixturing or feeding devices). Servos utilizing data from sensing the external environment require more complex processing of the signal but yield much more flexible systems.

The controller, in addition to controlling the manipulator motion, often also serves as an interface to the outside world—coordinating the robot's motions with machines and assembly lines and turning on and off machines that it is operating.

16-4.1. Pick-and-Place Robots

The simplest controllers are nonservo (open-loop) devices that rely on sequencers and mechanical stops to control the manipulator end-point positions along each of the axes. These limited-sequence robots are also referred to as pick-and-place, bang-bang, or fixed-stop robots. These robots, usually pneumatic, have no provision for trajectory control between the end points.

16-4.2. Programmable Robots

Programmable robots are servo-controlled robots of two basic types: point-to-point and continuous path.

Point-to-point robots are directed by a programmable controller that memorizes a sequence of arm and end-effector positions. Hundreds of points may be memorized. The robot moves in a series of steps from one memorized point to another under servo control using internal joint sensors for feedback. Because of the servos, trajectory control between the memorized points is possible and relatively smooth motions can be achieved.

Continuous-path robots do not depend on a series of intermediate points to generate a trajectory, but duplicate during the playback process the continuous motions recorded during the teaching process.* Thus these robots are used for painting, arc welding, and other processes requiring smooth, continuous motions.

---

*In reality, the internal form of the path is a set of closely spaced intermediate points.

16-4.3. Computerized Robots

Computers are being employed increasingly to control today's robots. Some of these robots can be treated as another computer NC (numerically controlled) tool. Many also support "teach pendants" for direct manual programming. Computer-controlled robots are capable of being programmed off-line using a high-level programming language and do not have to rely on being physically taught.

16-4.4. Sensory Robots

These are computerized robots that interface with the outside world through external sensing such as sight or touch. These "intelligent robots" are capable of adapting to a variety of conditions by changing their actions based on relating the information sensed to their goals or preprogrammed decision points.

16-4.5. Robots as Part of a Flexible Manufacturing System

A *flexible manufacturing system* (FMS) is a programmable batch-processing (small-lot manufacturing) arrangement that contains programmable machine tools and transfer devices all under central computer control. When robots are used in such a system, the robot controller may itself be just the bottom layer of a hierarchical control system for the FMS [see Toepperwein et al. (1980, pp. 96–105) or Albus (1981a, pp. 261–280)].

17

SENSOR-CONTROLLED ROBOTS

Sensor-controlled robots are the first step on the path to truly intelligent robots—robots that can determine their own actions based on their perception and planning abilities.

The simplest versions of sensor-controlled robots utilize contact switching for stopping the arms and for opening and closing grippers. More sophisticated robots are now beginning to use touch, force, and torque sensing. Forces and torques can be derived from stress measurements or from internal signals (e.g., derived from back electromotive force when using dc motors). Tactile information may be derived from deflection-induced resistance variations of a special sensing surface, or by various other ingenious approaches.

Work at several major robotic laboratories in the United States and abroad on "artificial skins" with embedded arrays of touch sensors opens the possibility of a robot eventually being able to determine position, orientation, and identity of parts by touch alone. This may be particularly important for difficult assembly operations (Kinnucan, 1981). Proximity and range sensing are also being considered.

Vision, the most complex of all the sensory modalities, is increasingly finding its way into robotic control.

18

PROGRAMMING A ROBOT

There are four primary methods for programming a robot:

1. *Physical setup*, in which the operator sets up programs by physically fixing stops, setting switches, arraying wires, and so on. This is characteristic of the simpler robots.
2. *Lead-through*, in which the operator leads the robot through the desired positions and locations by means of a remote "teach box." (These points are recorded and used to generate the robot trajectory during operation.)
3. *Walk-through*, where the robot arm is physically manipulated through the desired motions (which are recorded and then played back by the robot control during operation).
4. *Writing a software program*, which is then executed when desired.

The emphasis in programming research today is on software programming (mostly in higher-level languages) of computer-controlled robots. Work on sensor-controlled manipulation is extending the scope for programmability. Interacting with the robot by means of software provides more flexibility than the other programming methods and allows for conditional actions or flexible adaptations. Various high-level robot programming languages such as VAL (Unimation) and AML (IBM) are now beginning to become available to aid in software generation. Table 18-1 provides examples of VAL and AML programs for placing a peg in a hole.

Simons (1980, p. 107) notes that software programs can be divided into two types: "In *explicit programming*, the user requires explicit instructions for every action the robot must take. In *world modelling*, the robot is more autonomous and can make decisions according to its knowledge. *World-modelling* systems, largely in the research stage, tend to require a considerable amount of computer power but are able to carry out complex tasks."

Table 18-2 provides information on the status and capabilities of current robot programming languages. In general, these languages provide for off-line textual programming of robot motions. They all provide for coordinated motions in tool-point coordinates and most provide for visual servoing. However, except for AL, in general these languages do not as yet provide the capabilities for world modeling required for truly intelligent robots.

TABLE 18-1 Examples of VAL and AML Programs for Placing a Peg in a Hole

VAL		AML
SETI	TRIES = 2	PICKUP: SUBR (PART_DATA, TRIES);
REMARK	If the hand closes to less than 100 mm, go to statement labelled 20.	MOVE(GRIPPER, DIAMTER(PART_DATA)+0.2);
10 GRASP	100, 20	MOVE(<1, 2, 3>, XYZ_POSITION(PART_DATA)+<0, 0, 1>);
REMARK	Otherwise continue at statement 30.	TRY_PICKUP(PART_DATA, TRIES);
GOTO	30	END;
REMARK	Open the fingers, displace down along world Z axis and try again.	TRY_PICKUP: SUBR(PART_DATA, TRIES);
20 OPENI	500	IF TRIES LT 1 THEN RETURN ('NO PART');
DRAW	0, 0, −200	DMOVE(3,−1.0);
SETI	TRIES = TRIES −1	IF GRASP(DIAMETER(PART_DATA)) = 'NO PART'
IF	TRIES GE 0 THEN 10	THEN TRY_PICKUP(PART_DATA, TRIES −1);
TYPE	NO PIN	END;
STOP		
		GRASP: SUBR(DIAMETER, F);
REMARK	Move 300mm above HOLE following a straight line.	FMONS: NEW APPLY ($MONITOR, PINCH_FORCE(F));
30 APPROS	HOLE, 300	MOVE(GRIPPER, 0, FMONS);
REMARK	Monitor signal line 3 and call procedure ENDIT to STOP the program.	RETURN (IF QPOSITION (GRIPPER) LE DIAMETER/2
		THEN 'NO PART'
REMARK	if the signal is activated during the next motion.	ELSE 'PART');
REACTI	3, ENDIT	END;
APPROS	HOLE, 200	
REMARK	Did not feel force, so continue to HOLE.	INSERT: SUBR (PART_DATA, HOLE);
MOVES	HOLE	FMONS: NEW APPLY ($MONITOR,
		TIP_FORCE(LANDING−FORCE));
		MOVE(<1, 2, 3>, HOLE+<0, 0, .25>);
		DMOVE(3, −1.0, FMONS);
		IF QMONITOR(FMONS) = 1
		THEN RETURN ('NO HOLE');
		MOVE(3, HOLE(3) + PART_LENGTH(PART_DATA));
		END;
		PART_IN_HOLE: SUBR (PART_DATA, HOLE);
		PICKUP (PART_DATA, 2.);
		INSERT (PART_DATA, HOLE);
		END;

Source: Lozano-Pérez (1983, pp. 833–834). Used by permission of the IEEE. Copyright © 1983 IEEE.

175

TABLE 18-2 Current Robot Programming Languages

Robot Language	Organization	Control Mode			Manipulation Type		Status		Capability		Comments
		Position	Force	Vision	Rectilinear	Jointed	Commercial	Research	Assembly	Control of Multiple Arms	
AL	Stanford University	X	X	X		X	X	X	X	X	World model capability
AML	IBM	X	X		X		X		X		Menu capability
HELP	GE	X	X		X		X		X	X	
JARS	JPL	X	X	X		X		X			Designed for visual and force servoing
MCL	McDonnell-Douglas	X				X	X			X	For off-line programming of robots from a CAD data base
RAIL	Automatix	X		X	X	X	X	X	X		Developed for visual inspection, assembly, and arc welding
RPL	SRI	X		X		X	X	X			
VAL	Unimation	X		X		X	X		X		For control of machines comprising a robot work cell

Source: Based on Gruver et al. (1983).

19

ROBOT VISION TECHNIQUES

In the industrial manufacturing arena, to circumvent the requirement that the work-piece be in a prescribed position and orientation (pose) for the robot to operate upon it, sensory systems can be employed. Vision provides perhaps the most flexible approach to avoid all the fixturing that would be required to achieve a fixed pose.

Figure 19-1 lists the desired functions of a machine vision system to achieve fully flexible sensor-controlled manipulation. Figure 19-2 indicates potential applications for such a system.

19-1. METHODS FOR DETERMINATION OF POSE

1. *Range* can be determined in four principal ways:
   a. Stereo
   b. Triangulation
   c. Active ranging (e.g., time of flight of light or sound)
   d. Using optical focusing
2. *Orientation* can be determined by:
   a. Observing the relationship of three (or more) known object points that are not all collinear in the viewing field, provided that the relative ranges connecting these points are known (approach difficult to implement but has been considered for assembly of components in space)
   b. Deriving surface normals using the intrinsic image concept (Barrow and Tenenbaum, 1981)

- Recognition of workpieces/assemblies and/or recognition of the stable state* where necessary.

- Determination of the position and orientation of workpieces/assemblies relative to a prescribed set of coordinate axes.

- Extraction and location of salient features of a workpiece/assembly to establish a spatial reference for visual servoing.

- In-process inspection—verification that a process has been or is being satisfactorily completed.

*One of the resting orientations that the workpiece can assume.

Figure 19-1 Desired functions of machine vision for sensor-controlled manipulation in industrial manufacturing. (From Rosen, 1978. Used by permission of SRI International, Menlo Park, CA.)

c. Deducing orientation based on the response of lighting on the object, utilizing known characteristics of the object (approach simplified if structured lighting such as sheets of light are employed)

19-2. FEATURE EXTRACTION

Although it is possible to operate directly on an analog output of a vidicon imager (by such approaches as optical correlation or template matching), it is more expedient (and programmable) to use solid-state imaging arrays, such as CCD or CID devices, which directly provide points (pixels) in the image.

These pixels are then digitized by intensity levels (usually 2 to 256) so as to be put in a convenient form for computer manipulation. From these data, edges and regions are determined. Edges are considered to occur when there is an intensity gradient exceeding a selected threshold. These gradients are derived from intensity differences between adjacent pixels (often using an averaging scheme to reduce noise effects). Regions are grown by annexing ("clustering") adjacent pixels that are close to the same intensity level. Various schemes (heuristics) exist to connect lines having gaps, smooth irregularities, and remove discontinuities in regions (all resulting from noise in the image). The result (which can be considered as a form of data compression) is a reduced image that consists of edges and regions. As indicated in Fig. 19-3, further data compression can be achieved by representing edges by sets of straight or curved lines (the ends and curvature of which are all that need to be known).

19-3. OBJECT RECOGNITION

Having determined edges and regions, it is possible using the methods of Section 19-1 to determine position and orientation. However, often it is necessary to determine which object is being observed from a collection of possible objects. This may be accomplished by:

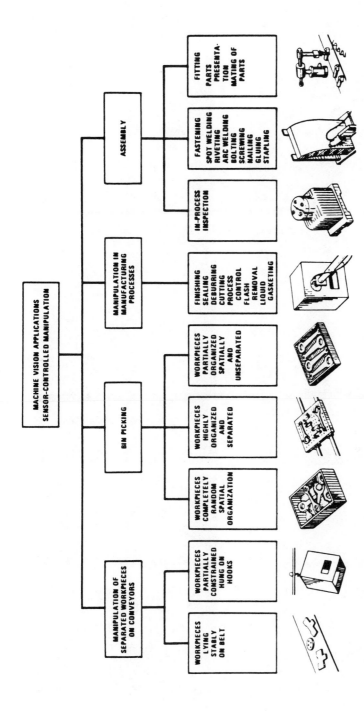

Figure 19-2   Block diagram of potential vision applications in robotics. [From Rosen, 1978, p. 5. (Illustrated version shown is from Saveriano, 1980, p. 15.)]   Used by permission of *Robotics Age* and SRI International, Menlo Park, CA.

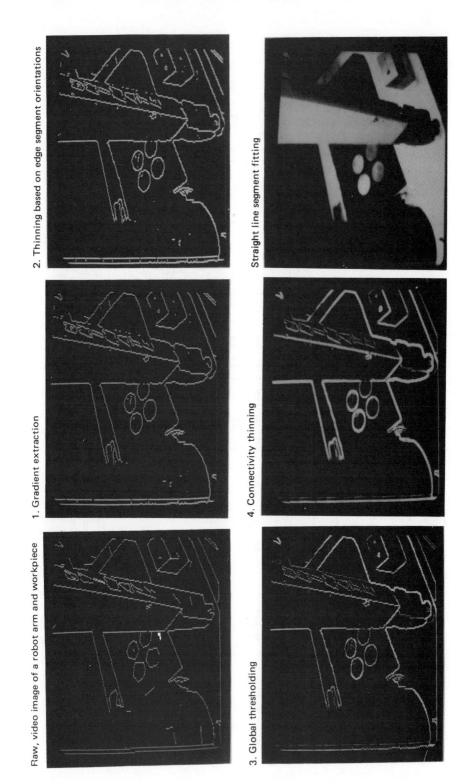

Raw, video image of a robot arm and workpiece

1. Gradient extraction

2. Thinning based on edge segment orientations

3. Global thresholding

4. Connectivity thinning

Straight line segment fitting

Figure 19-3 Edge extraction using the JPL IMFEX real-time pipeline processor (steps 1–4). (Courtesy of NASA Jet Propulsion Laboratory.)

1. *Template matching:* using stored images in the computer. This approach is valid only if a small number of stable states suffices.
2. *Use of edge and region statistics:* curvature of edges, area moments of inertia, ratio of square of perimeter to area, number and/or location of holes or notches, and so on.

19-4. FUTURE TECHNIQUES

In the future we may see:

- Greater use of direct range measurements e.g., scanning laser radar, or other optical devices using phase or intensity measurements
- Parallel processing to obtain feature extraction in real time
- Improved methods for determining edges and regions
- Use of optical markings for recognition, for optical code reading, or for determination of pose
- Use of intrinsic image analysis* to determine surface orientations, depth, relationship of foreground to background, and texture
- Artificial intelligence-based scene analysis for less structured situations, as indicated in Chapter 8.

---

*This approach, advocated by Barrow and Tenenbaum (1981, pp. 581–582), is discussed in Chapter 8. It goes beyond two-dimensional representations and regions to a representation proposed by Marr and Nishihara (1978) of MIT, called the *2.5D sketch*, consisting of surface distances and orientations, useful for finding three-dimensional surfaces and volumes for interpretation.

20

KINEMATICS AND FLEXURE

20-1. COORDINATE TRANSFORMATIONS

If the arm is programmed by walking-through, it is necessary only to record the simultaneous joint positions during the walk-through. If the arm is led-through, in more sophisticated systems that can operate in tool-point, world, or other coordinate systems it is necessary to do the needed coordinate transformations for each joint to obtain the desired end-effector orientation and position. This is particularly true when the arm is being programmed by writing software.

When the arm is controlled via feedback from a vision or other sensor, it is necessary to do coordinate transformations to convert the camera or hand coordinates to base or computer coordinates. For vision (or other sensor) control, the sensor interpretation and coordinate transformations need to be done in real time. As the mathematics for exact transformations requires time-consuming calculations, approximate simplifications have been developed to speed the calculations. In some systems, individual microprocessors are used to process the kinematics (the motion) at each joint.

Arms are often designed so that the wrist motions are mechanically constrained to cylindrical or rectilinear coordinates to simplify control and reduce coordinate transformation requirements.

20-2. TRAJECTORY SELECTION

In moving the end effector from one position (or point) to another, various schemes are possible for choosing the motions for moving the joints. Approaches for articulated arms include:

- Variations from preprogrammed paths.
- The "dominant-axis" approach: some joints may be allowed to adapt, whereas others follow fixed movements.
- All joint motions move proportionately relative to the slowest joint, so that all joint motions terminate at the same time.

Considerations of this sort become particularly important for off-line or real-time programming of computer-controlled robots.

20-3. ARM FLEXURE

At present, all robot arms are made relatively stiff so that structural flexibility during gravity or dynamic loading is minimized (to promote repeatability). However, this stiffness requirement usually results in heavy arms with high-power actuators, lowered carrying capacity, and reduced speed. Research is now under way on the control of structurally flexible arms to help overcome these limitations, and in active control to control flexure for achieving on-line precision under load as during assembly or drilling operations. The 15-m-long Remote Manipulator System (RMS) for the Space Shuttle is probably the most extreme case of a structurally flexible manipulator used to date.

21

ROBOT MOBILITY

At present, most robots in industrial applications are fixed in place or ride along rails, guideways, or conveyors. Simple robot carts for transporting materials are now being used in industrial and commercial applications. Robot vehicles using wheels and treads have been researched by NASA for planetary exploration purposes. In addition to research, wheeled robots, such as the Heathkit HERO robot, are now being utilized for educational and publicity purposes. In the future, we can expect wheeled robots in a variety of applications in areas such as the service industries and in applications in the home.

Crawling robots that use their manipulator arms to grasp handholds are being explored by NASA for use in constructing large structures in space. Walking machines are still in the experimental stage, with research going on at Ohio State University, Carnegie-Mellon University, Tokyo Institute of Technology, and in the Eastern block countries.* Engelberger (1980, p. 63) states that "practical robot devices with legs could have advantages in such applications as arctic transport, mining, agriculture, forestry, fire-fighting, explosive ordnance disposal, unmanned ocean-floor surveys, and planetary exploration."

Walking machines are now beginning to receive serious consideration. Ohio State University is building a full-scale six-legged person-carrying supervisory-controlled machine for the army. At the 1983 International Symposium on Robotics, Odetics, Inc., unveiled a six-legged teleoperated walking machine (Fig. 21-1) dubbed a "functionoid." The Soviets are also reported to be very active in the development of walking machines.

*A good indication of current research on walking machines is given by *Robotics Research*, Special Issue on Legged Locomotion, Vol. 3, No. 2, Summer 1984.

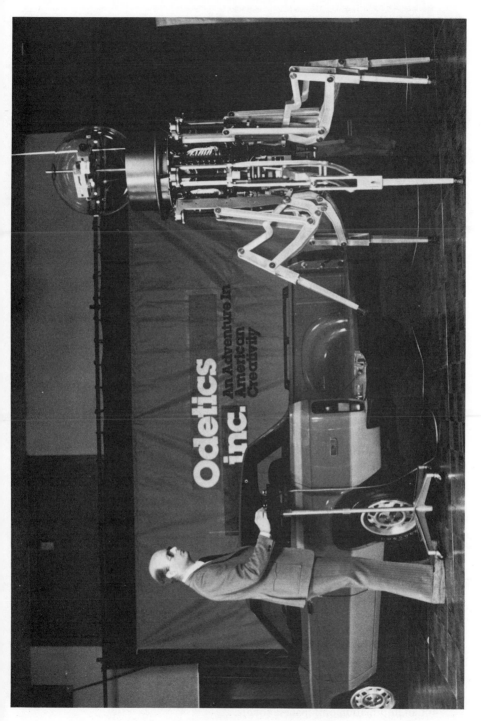

Figure 21-1 Mobile robotics: ODEX 1 (shown stepping into the back of a truck under tele-operator control). Used by permission of Odetics, Inc., Anaheim, CA.

22

COMPETITIVE SYSTEMS AND
THE STATE OF THE ART

22-1. ROBOTS

Engelberger (1980, p. 117) indicates that many of the attributes required to make industrial robots practical in an industrial environment are available today. These (see Fig. 22-1) include good positioning repeatability, reliability, flexibility, easy programmability, and adequate load capacity, all at an affordable price.

The rapid growth of industrial robots (roughly 35% per year) and the entry of new manufacturers and new products makes it inappropriate to try to list or even summarize all the robots currently available. However, the following are good sources of information to keep current with available robot products:

- *Robotics Age*, a bimonthly magazine popularizing robots, published by Robotic Age, Inc., 174 Concord St., Peterborough, NH 03458.
- *Robotics Industry Directory*, published yearly by Technical Database Corporation, P.O. Box 720, Conroe, TX 77305. This directory lists complete information on more than 100 different models of currently available robots, robot components, and peripherals. It also provides information on robot distributors, robot research institutes, and consultants in the robot field. A periodic robotics update is also available.
- *Specifications and Applications of Industrial Robots in Japan*, published by Japan Industrial Robot Association (JIRA), 3-5-8 Shiba Koen, Minato-ku, Tokyo, Japan. (*ROBOT*, a periodical, is also available.) JIRA is a good source for all information on Japanese robots.
- *Robotics Today*, a periodical on industrial robotics, published by the Society

1. Work space command with six infinitely controllable articulations between the robot base and its hand extremity.

2. Teach and playback facilities—realizing fast, instinctive programming.

3. Local and library memories of any practical size desired.

4. Random program selection possible by external stimuli.

5. Positioning accuracy repeatable to within 0.3 mm.

6. Weight handling capability up to 150 kilos.

7. Point-to-point control and continuous path control, possibly intermixed in one robot.

8. Synchronization with moving workpieces.

9. Interface allowing compatibility with a computer.

10. Palletizing and depalletizing capability.

11. High reliability—with not less than 400 hours MTBF.

12. All the capabilities available for a price that allows purchase and operation within the traditionally accepted rules for economic justification of any new equipment.

Figure 22-1   Robot attributes now commercially available. (Adapted, by permission of the publisher, from *Robots in Practice* by Joseph F. Engleberger, p. 117, © 1980 by Joseph F. Engleberger. Published by Amacom, a division of American Management Associations, New York. All rights reserved.

of Manufacturing Engineers in cooperation with the Robot Institute of America,*One SME Drive, P.O. Box 930, Dearborn, MI 48128.

- *The Industrial Robot*, a magazine with a European orientation toward industrial robots, published by IFS Ltd., 35-39 High St., Bedford MK42 7BT, England.

- *Robotica*, an international journal of information, education, and research in robotics and artificial intelligence, published by Cambridge University Press, 32 E. 57th St., New York, NY 10022.

Other robotics research journals are now available from MIT, IEEE, and other research and engineering organizations.

In addition, various robotic newsletters (such as *Robot News International*, also available from IFS Ltd.; *Industrial Robots International*, available from Technical Insights, Inc., P.O. Box 1304, Ft. Lee, NJ 07024; *Robot Insider*, available from Fairchild Publications, 11 E. Adams St., Chicago, IL 60603; and the *Robotics Newsletter*, available from Bache, 100 Gold St., New York, NY 10038) offering information on new robot products and investments have begun to proliferate. *Datamation, Byte, Business Week, Time,* and other technical and popular magazines also have had feature articles on robotics.

At the end of 1981, there were some 50 U.S.-based firms marketing or preparing to market robots (Conigliaro, 1981), the leaders being Unimation Inc., now a subsidiary of Westinghouse [estimated to have had one-third of the roughly $215

---

*Now called the Robotic Industries Association.

million (3075 units) U.S. market in 1982] and Cincinnati Milacron (estimated to have had one-fourth of the U.S. market in 1982). Table 22-1 provides an indication of the past distribution of robots by manufacturer in the United States.

In addition, several of the large U.S. automobile, electrical, and computer firms have been building robots for internal use and have now entered the commercial robot market. These include giants such as IBM, General Electric, General Motors, Texas Instruments, Allegheny International, United Technology, and Textron. By April 1983, there were over 100 U.S.-based firms marketing robots—a doubling in two years, with the 1983 market estimated at $250 million. Some 40 vision companies were also in operation. It might even be that the machine perception business may be larger than the robotics business.

Congliaro (1983) indicates that outstanding startup companies have been:

- Advanced Robotics Corporation
- American Robot Corporation
- Automatix
- Control Automation
- Intelledex
- Machine Intelligence Corporation
- Nova Robotics

In the process, market shares have changed dramatically. It now appears that it is probably too late for new companies to enter the robotics industry easily. The successful companies appear to be those that have found a relatively unique niche, such as spray painting or arc welding, or are acting as a "value-added" systems house that does more than sell a robot.

Consolidation and combination may be the order of the day. Westinghouse has acquired Unimation, and many of the robotics companies have made agreements with Japanese robot manufacturers to buy their high-quality (and lower-cost)

TABLE 22-1   Estimate, by Manufacturer, of U.S. Robots Installed (End of 1981)

Manufacturer	Number of Robots
Unimation	2500
Cincinnati Milacron	1200
Prab	1000
Copperweld	735
DeVilbiss	140
Asea	85
Others	450
	6000

Source: Electric Power Research Institute (1984). Copyright © 1984, Electric Power Research Institute. Reprinted with permission.

manipulators and then add value in terms of better computer controls, sensors, software, and pre- and post-installation support and other auxiliary equipment. Technical and sales agreements between U.S. and Japanese robot companies are now nearly universal. Agreements between GM and Fanuc for large numbers of robots will have a major impact on the future distribution of robots in the United States.

At the end of 1981 Japan had more than 130 manufacturers producing robots, many for internal use. Table 22-2 presents the robot production by major Japanese manufacturers for the year 1982. Of the estimates by JIRA of roughly 24,000 robots valued at $590 million produced in Japan in 1982, only 14% were exported.

In 1981 approximately 4000 robots were sold in Europe. European manufacturers having a major share of this market include ASEA in Sweden, Olivetti and DEA in Italy, VW, Kuka, and Seamens in Germany, and Renault in France.

The Eastern block countries have been moving in the direction of robots to make up for their serious labor shortages. *Business Week* (August 17, 1981) reported that the Soviets plan to manufacture large numbers of robots to counter not only the shortage of skilled labor, but also absenteeism associated with drinking. Poland also has been aggressively pursuing robot design and development.

22-2. COMMERCIALLY AVAILABLE VISION SYSTEMS FOR ROBOTS

Present commercially available vision systems for use with robots are basically devices that generate and analyze two-dimensional images. These systems recognize objects by their silhouettes. Principal manufacturers of sophisticated computer

TABLE 22-2   Approximate Value of Robots Sold
in 1982 by the Six Largest Japanese Robot
Manufacturers[a] (millions)

Matsushita Electric Industries	$ 54
Hitachi	33
Kawasaki Heavy	31
Yaskawa Electric	28
Fanuc	25
Dainechi Kiko	19
	$190

[a] Combined robot sales of these six manufacturers represent 32% of total 1982 Japanese robot sales of $590 million.

Source: Aron (1983), p. 53. Used by permission of Paul H. Aron, Vice Chairman of the Board of Directors, Daiwa Securities America Inc., New York, N.Y. and Professor of International Business at the Graduate School of Business Administration of New York University, NYC.

vision systems are indicated in Chapter 8. The following early commercial machine vision systems is indicative of some of the vision systems now available.

22-2.1. M.I.C. Vision Module
(Machine Intelligence Corp., Sunnyvale, California)

The Vision Module is based on prototype software and hardware developed at SRI under NSF and industrial affiliate support (see Gleason and Agin, 1979). The Vision Module consists of a solid-state TV camera, a preprocessor interface, and a LSI-11 microcomputer. The vision software, stored in the LSI-11 memory, includes the entire library of vision subroutines previously developed at SRI, several application programs (such as training by showing and execution of part recognition), and newly developed capabilities. The ability to develop application programs by calling the vision-library subroutines makes the Vision Module a general-purpose vision system. The Vision Module is intended for a variety of industrial vision applications, such as in-process inspection, sensor-controlled manipulation, and visual servoing.

The Vision Module can recognize parts on the basis of their size and shape regardless of their position or orientation. The part (lying in one of its stable states previously used to train the module) is recognized on the basis of a set of shape descriptions, including area, perimeter, number of holes, length of major and minor bounding ellipse axes, first and second moments of area, and various algebraic combinations of these.

The Vision Module operates by appropriately thresholding the image in the preprocessor to achieve binary silhouettes of the objects in the scene. A frame buffer stores the binary image (consisting of zeros and ones) to achieve timing independence between the camera and the controlling microcomputer. Each line is sequentially scanned and "run-length encoded" (by recording edge points where the pixels change from zero to one, or vice versa). Each run-length segment on a line is matched against segments of the previous line to determine their overlapping relationships. Using these relationships, the program traces the appearance and disappearance of regions as the image is processed from top to bottom. Only one line at a time of image data needs to be stored in the computer.

At the heart of this vision processing approach are connectivity analysis routines which segment the image into contiguous regions (called *blobs*) even in the presence of noise. During the connectivity analysis, sequences of perimeter points are extracted. Using this blob boundary information, other desired parameters, such as area, area moments, shape, position, and orientation of the blob, can be computed.

The system is trained by showing the object to the TV camera, resulting in all potentially useful shape descriptions being automatically calculated and stored. Then the same object is moved and photographed several times and the mean and variance of these descriptors calculated and stored. This resulting set of means and variances can be used as a "prototype" for that object or class of objects. New objects can then be recognized in either of two ways: (1) by using a normalized nearest-neighbor method, where the best fit of the features of the new object to the set of prototypes is selected; and (2) using a binary decision tree (which can be built by the Vision Module software upon command), recognition is done by measuring one feature at a time, and at each stage subdividing the recognition problem into two smaller problems until a single choice remains. The time to read in, process, and implement recognition is about 1 second.

Complete vision systems cost from $35,000 to $45,000. M.I.C. also markets the VS-110 Machine Vision System version for real-time inspection tasks. M.I.C. is now placing increased emphasis on turnkey systems, visual-inspection workstations, and improved human-machine interfaces. The problems of resolution, speed, and cost are also being vigorously attacked.

22-2.2 The Automatix Autovision System (Automatix, Billerica, Massachusetts)

The Autovision System (Rheinhold and Vanderbrug, 1980) also uses the SRI algorithms as its main base for vision processing and in many ways is similar to the M.I.C. system. However, the Autovision System's preprocessor (Scheinman, 1981) has a pixel memory capable of storing several images at once, or several bits of gray-level information, or any combination. The versatility of the system comes from the system's powerful AI-32 control processor. This control processor with its 32-bit internal design and megabyte memory address space can operate in parallel with the preprocessor and has the ability to interface with multiple preprocessors and various other input and output devices. The control processor operates under an Automatix-developed operating system. Most of the system software is factory programmed in Pascal, except where speed or compactness is so critical that assembly language or microprogramming is used.

The primary interface for the user is via the Automatix application language, RAIL, which is designed for the non-computer-oriented user. This powerful and flexible robot programming language makes it possible to sequence the operations of the Autovision system so that the desired task can be performed.

Automatix's primary focus is now on manufacturing vision-controlled robots for such applications as welding, assembly, and inspection. Automatix is also exploring the use of force and other sensor modalities.

22-2.3. KEYSIGHT AND CONSIGHT

At General Motors, several different computer vision systems are now operational (Rossol, 1981). KEYSIGHT is an example of an early system for inspection tasks in automobile manufacturing. It uses a low-resolution solid-state camera to inspect for the presence or absence of valve spring cap keys on automobile engines. Ingeniously applying fairly standard computer vision algorithms for edge finding, and so on, the system locates the center of the valve spring assembly and then, based on the keys being brighter than the holes they fit over, determines if the keys are present.

More interesting is CONSIGHT (see Figure B22-1), an example of a robot vision system that uses structured light to detect the silhouette of passing objects. Using a robot arm, CONSIGHT transfers a moving stream of nontouching, randomly oriented parts from a moving conveyor belt to a predetermined location.

Structured light provides a means to deal with situations that require both high spatial resolution and fast response. It allows part geometry to be sensed directly, as opposed to being inferred through gray-level picture processing.

CONSIGHT is logically partitioned into independent vision, robot, and monitor subsystems. When the vision subsystem has seen the entire object, it sends to the monitor the object's position and belt position reference value. The robot subsystem then executes a previously "taught" robot program to transfer the part from the belt to a fixed position.

CONSIGHT is now commercially available, having been licensed to Unimation, Cincinnati Milacron, M.I.C., and Automatix.

CONSIGHT, developed by General Motors, is a visually guided robot system that locates and transfers randomly positioned parts from a moving conveyor belt. The vision system employs a linear array camera that images a narrow strip across the belt perpendicular to the belt's direction of motion. Since the belt is moving, it is possible to build a conventional two-dimensional image of passing parts by collecting a sequence of these image strips. For each object detected, a small number of numerical descriptions is automatically extracted for part classification and position determination.

Structured light* consisting of a narrow and intense line of light is projected across the belt surface. The line camera is positioned so as to image the target line across the belt. When an object passes into the beam, the light is intercepted before it reaches the belt surface (Fig. B22-2). When viewed from above, the line appears deflected from its target whenever a part is passing on the belt. Therefore, wherever the camera sees brightness, it is viewing the unobstructed belt surface; wherever the camera sees darkness, it is viewing the passing part.

Unfortunately, a shadowing effect causes the object to block the light before it actually reaches the images line, thus distorting the part image. The solution is to use two light sources (one before the strip and one after the strip) directed at the same strip across the belt. When the first light source is prematurely interrupted, the second will normally not be. By using multiple light sources and by adjusting the angle of incidence appropriately, the problem is essentially eliminated.

*Structured light can also be used in other schemes such as determining object shape, orientation, and location based on the perceived size, location, and distortion of the projected lines.

Figure B22-1 CONSIGHT: Example of a special-purpose vision system using structured light. From Rossol (1982) pp. 4-5. Used by permission of General Motors Research Laboratories; Warren, Michigan.

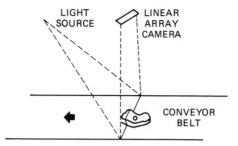

Figure B22-2 Camera is positioned to image target line across the belt. From Rossol (1982). Used by permission of General Motors Research Laboratories, Warren, Michigan.

23

RESEARCH REQUIRED

Robots have come a long way since the first commercially available robot was sold by Planet in 1959. However, much remains to be done to make them more versatile, productive, cost-effective, and lighter.

23-1. MANIPULATOR AND ACTUATOR DESIGN

To achieve the higher speeds, flexibility, accuracy, efficiency, and dexterity that is desired in future robots, additional research in manipulator arms are needed. For large manipulators and precision work, joint and arm structural flexibilities can be a problem. Therefore, stiff, but lightweight arms are desirable to minimize this problem and to aid in increasing speed. However, as variable compliance in different directions is often desirable to provide the "give" needed for inserting parts and other assembly operations, unique design approaches may be required.

At present, actuators are one of the more problem-prone areas of robots. In addition, actuator efficiencies are in the order of one-tenth that of human musculature. Smaller, more reliable and efficient actuators and drive mechanisms, with high load-to-weight ratios, are needed for the high-speed, lighter-weight robots of the future.

For greater dexterity and less intrusion into the work zone, it is also necessary to understand the mutual interdependence of end effector and manipulator kinematics. Also, research on micro-manipulators is needed for applications to microelectronics and biology.

A unified theory is needed on how to design an arm for smooth universal movements, satisfaction of functional requirements, and to provide the required load capacities throughout its motion spectrum. A standardized approach to performance measurement and evaluation is required to enable comparison of robots not only in such factors as working volume, load capacity, and reliability, but also in aspects such as speed, smoothness, and accuracy throughout their operating range.

To help determine manipulator accuracy (now a limiting factor in off-line programming of open-loop manipulator operations), a universal stochastic model of mechanical errors of manipulators needs to be devised.

23-2. END EFFECTORS

Umetani (1981, p. 110) states that "the most difficult barrier in industrial robot design . . . is the design of dexterity in hands and fingers." Thus much work is needed to devise designs for general-purpose hands (and the means to control them) such that the whole arm need not move to provide the desired dexterity. Because general-purpose hands are not yet available, multiple end effectors may be required for a single application. This, in turn, requires some type of quick-change wrist design.

23-3. CONTROL

Kelly and Huston (1980) indicate that further research is needed on the basic robot problem: Given the desired external or absolute displacement, velocity, and acceleration of the hand, determine the optimal physical and internal motion parameters. Birk and Kelley (1981) state that procedures have been recently developed for efficiently obtaining and solving governing dynamic equations for robotic systems with flexible links and joints. However, optimizing the arm motion still presents very serious problems of convergence, uniqueness, and computational complexity (Kelly and Huston, 1980).

Although the repeatability of arm positioning via open-loop (no external sensory feedback) robot control has been good, accuracy for initially positioning the arm via open-loop control has been poor. In addition, no efficient universal means of robot calibration has yet been determined. Thus it is important that more work be done on improving open-loop control accuracy or developing smarter closed-loop controllers. Accuracy degradation due to flexure can be a particularly troublesome problem when trying to perform precision operations under load, such as drilling without the aid of jigs.

An important area of control not yet solved is control of structurally flexible manipulators. This problem, important to high-speed manipulation with manipulators having high load-to-weight ratios, may require real-time simultaneous management of a large number of command and control variables. An associated problem with such manipulators may be the large changes in moments of inertia as the manipulator is extended and a load applied. Another problem associated with high-speed manipulation, requiring additional research, is the generation of time-optimal trajectories.

The control of redundant-axis manipulators, needed to provide additional manuever capability, is an area in which little is known about ideal solutions.

Design of control systems for fault tolerance will become particularly important as robots become more autonomous.

Finally, lack of suitable strategies for control of multifingered hands is still one of the principal factors retarding design of dexterous hands.

23-4. SENSORY-CONTROLLED MANIPULATION

One of the most desirable features for improving robot flexibility is sensory capability. Vision, the most popular U.S. robot research area, appears to be the most desirable sense for present and future IRs. Next in importance is touch and force sensing.

To take full advantage of sophisticated sensory data, real-time control using sensors is needed. Hierarchical control (Albus, 1981a) may be the appropriate means to handle such data, although control logic systems such as pattern reference feedback logic may be appropriate in certain contexts.

In addition to vision and tactile sensing, dynamic control using force feedback is an important area of research. Also of increasing importance as robots become more autonomous is proximity sensing as used before final contact and to avoid obstacles. Other sensors will increasingly have roles in inspection and in determining the type of material being observed. Research is needed in the use of models in sensory processing and interpretation.

It is also necessary to explore the economic trade-offs between the use of sensors and imposing more structure on the robot environment. As sensory control is perfected, the trade-off will undoubtedly shift toward the direction of using sensors.

23.5. VISION

Vision becomes more important as the environment becomes less structured. Therefore, vision is needed more for firefighting, household servants, robot explorers, and so on, than for IRs. Vision can also reduce manipulator accuracy requirements by making real-time adjustments practical.

At the current state of the art, additional research in recognition and geometric representation is needed. As greater capability is desired, three-dimensional vision combined with world models and spatial reasoning will become important. Current three-dimensional research in vision systems center on the use of range finding, structured light, and binocular vision systems. Further out, research is required in the analysis of image sequences (optical flow, shape from motion), intrinsic image analysis (determining shape from a single image by disambiguating the response to light), and improved methods to recognize objects in clutter, such as picking an object out of a bin.

One important research area in vision is how to improve the speed of visual processing. This includes research in computational elements structured for visual processing, which includes parallel-processing CPUs, and special chips for edge finding, Fourier analysis, and so on. Even more important is fundamental algorithm development.

23-6. TACTILE AND FORCE SENSING

What is needed is better use of contact-sensing data. This includes better resolution in touch and force sensing, as well as better control strategies which make use of such information. An important result could be the ability to recognize parts and to determine their relative position and orientation by such methods.

23-7. PROGRAMMING A ROBOT

Programming by teaching is not practical for small lot production, particularly when sensory interaction is involved. Thus improved techniques are needed for generating robot control data. These include high-level programming languages, software debugging tools, world models, and the ability to directly utilize CAD data bases and sensory inputs. Robot simulation tools can aid in off-line programming. Also needed are better geometric modeling systems. As robot usage advances, a need arises for programs to orchestrate multiple arms and multiple sensors and for systems with distributed control.

23-8. INTELLIGENCE

Plantier et al. (1981, p. 63) assert that "Artificial Intelligence is the technological area which needs most to be developed and mastered to accelerate robot evolution." Birk and Kelley (1981) indicate that an intelligent robot is one capable of:

- Receiving communication
- Understanding its environment by the use of models
- Formulating plans
- Executing plans
- Monitoring its operation

Thus research in all these areas is important.

23-8.1. Communication

Use of a natural language interface or computer graphics or even voice commands would all be helpful in simplifying communication with a robot. In addition, intelligent aids for robot users to help plan and develop robot usage are also needed.

23-8.2. Knowledge Representation

A key ingredient in advanced robotics is knowledge representation. Sophisticated internal models of the environment are needed to generate expectations, to interpret sensory data, and for use in automatic generation of robot control data (such as for real-time decision making).

Representations are also needed to enable qualitative reasoning about physics and space for use in planning and problem solving. Parallel processing will be needed to make real-time modeling a reality.

23-8.3. Planning

Automatic problem solving or planning is one of the keys to future intelligent robots. Planning is even needed for acquiring information from sensors. Research is needed on ways to use large amounts of domain-specific knowledge to guide planning and responses to recognized situations.

One of the difficult but important areas of research needed is in developing plans that deal with time. Such research is exemplified by the work of Malik and Binford (1983) and Allen and Koonmen (1983).

For certain tasks, the software for handling contingencies is becoming excessive. Thus means for automatic monitoring and correction are needed. In addition, at a higher level, techniques for monitoring a plan execution and replanning as required is also in need of research. It would also be useful to be able to store successful plans in categories so that they can be reused. In general, it is desirable that the system be capable of adaptation and learning.

Automatic assembly in non-highly structured environments is a key research area. Development of kinematic strategies for assembly is needed. Assembly could be aided by model-based generation of programs that use local sensory information in assembly tasks. Research is also needed for planning and orchestrating the actions of cooperating robot systems.

Intelligent controls can be thought of as coupling control theoretic approaches with advanced AI methodologies. To advance this area, better understanding of modular hierarchical structures is needed as well as how to generate hierarchical plans associated with such structures.

To make intelligent robots really useful, it is necessary to develop methods that will enable them to function in real time (not having to pause for computation) with realistic computers.

23-9. ROBOT MOBILITY

For industrial robots tending machines, furnaces, and so on, the robots are often in use only a fraction of the time, the remainder being in a wait mode. To make better use of the robots, machines are frequently arranged so that one robot can tend several machines. Sometimes this involves moving the robot (motion along tracks being the usual means). Locomotion is needed to further extend the working range and for use in hazardous environments and in remote locations. In many cases, a truly mobile robot that can be used where needed would be preferred. Mobile robots would be particularly advantageous in environments less structured than that found in factories, such as in the construction of ships, buildings, and roads, in mining, and in undersea or space use.

New mechanical mechanisms, other than wheels, are needed for mobility over uneven terrain. However, mobility systems tend to amplify compliance and vibration problems (which can be minimized by careful design and the use of sensor-driven control). Active suspensions are needed for uneven terrains, climbing, and so on. For legged vehicles, the key problems are stability, control, and optimal motion synthesis. In walking machines very little is known about the stability of various gaits. Advances are needed in the hierarchical control of multibody systems. Efficient algorithms need to be developed for solving the governing dynamic equations.

If the mobile robot is to be supervisory-controlled, optimal partitioning of the task between the operator and machine becomes an important research area.

23-10. MODULARITY

Further work is needed not only to develop modular manipulators, but also to enable integrated modular systems that can consist of a variety of robots, tools, sensors, and computers all connected together. To do this, standardized interfaces are needed. It will be necessary to partition the control problem into modular components and to develop interface standards for communication between these components. An important goal is to be able to write an off-line program for one robot and to use it with a differently designed robot.

The 1980 NBS/Air Force ICAM Workshop on Robot Interfaces *Proceedings* (1981) considered the following items for standardization.

- Simple sensor interface between simple peripheral devices and a robot control system

- Wrist interface between the robot wrist and the end effector
- Complex sensor interface that covers vision, complex touch, and other such sensors
- Common robot control interface, providing robot-independent trajectory descriptions
- Robot programming languages
- Integration of robots into robot systems for automated manufacturing

It was concluded that both the simple sensor and wrist interfaces are ready for standardization, but that it was too early in the development of the other areas to begin a standards effort at that time.

24

PLAYERS, RESEARCH, AND FUNDING

24-1. IN JAPAN

Robot-related research in Japan can be placed into three categories: research done by universities and research institutions, by industry, and by special government-sponsored task forces. A 1980 JIRA survey found that a total of 88 research projects were then under way at universities and public and private research institutions. However, unlike the United States, there were no large-scale research efforts at the universities or technology colleges.

Areas of emphasis in research projects at Japanese universities and research institutions (Hasegawa, 1981, pp. 162-163) include the following:

- Increasing robot speed
- Coordination of multiple arms
- Dexterous robot hands with tactile recognition capability for automatic assembly
- Sensory development for intelligent robots
- Robot locomotion
- Robot control and application software development

Although the research activities at universities and public research institutions has increased substantially since 1980, it is clear that the overwhelming source of robotics research and development expenditures is private enterprises (Aron, 1983).

24-2. IN THE UNITED STATES

Robot research in the United States is aimed primarily in the direction of making robots more intelligent and versatile, with vision research being given a high priority. Albus (1981b, p. 7) states: "Much remains to be done in sensor technology to improve the performance, reliability and cost effectiveness of all types of sensory transducers. Even more remains to be done in improving the speed and sophistication of sensory processing algorithms and special purpose hardware for recognizing features and analyzing patterns in space and time."

Robotic research efforts at U.S. universities are centered on greater dexterity, improved sensory perception, higher-order robot programming languages, increased robot intelligence, and computerized manufacturing and assembly. The principal universities involved in robotics research are Carnegie-Mellon, MIT, Stanford University, Rhode Island University, Florida University, and the University of Michigan, with their combined 1983 funding being in the order of $10 to $15 million. The focus of robotics research in the U.S. nonprofit research laboratories (Charles Stark Draper Labs. and SRI) is on manufacturing, with emphasis on computerized planning, factory management, assembly, and inspection.

The emphasis in robotic research efforts in commercial U.S. companies is on manufacturing, assembly, and computer vision for part recognition and inspection. Research and development in some of the U.S. companies specializing in sophisticated computer vision stresses vision for inspection and robot control and on simplified command capability.

Focus of research activities at the principal U.S. robot manufacturers is on advanced control systems, improved programming techniques, and mechanical design.

Although there is no central focus for robotics in the U.S. government, several government agencies are involved in robotics research and a number of others are looking into applications.

Table 24-1 [roughly reflecting FY (fiscal year) 82/FY 83 funding] provides an indication of robotics research performed within the U.S. government. In general, these activities can be viewed as exploratory research into robotic applications.

Robotics research supported by the U.S. government is funded primarily by DARPA, the National Science Foundation (NSF), and the Office of Naval Research (ONR). NSF primarily supports basic research, DARPA and ONR support a mixture of basic and applied research, and the Veterans Administration focuses on potential applications for the handicapped.

DARPA, the Air Force, and the Army also support robotics applications development in the manufacturing and rework areas and in battlefield robotics.

The various U.S. military services are now undertaking robotic research efforts that could lead to replacing human beings, or leveraging their capabilities, for dangerous jobs or missions and for increased productivity.

It is estimated that overall U.S. government support for robotics research and development in the FY 1982/FY 1983 period was approximately $20 million per year.

TABLE 24-1 In-House Government Robotic Research (FY 82/FY 83)

Organization	Research Activities

National Bureau of Standards
$2 million/year
	Standards
	  Robot interface standards
	  Robot performance measures
	  Programming language standards for
	    Robot systems
	    Integrated computer-aided manufacturing
	Advanced concepts for
	  Sensory-interactive control systems
	  Modular distributed systems
	  Sensor interfaces to the control systems of robots and
	    machine tools
	  Vision systems
	Focused Application: Automated manufacturing research
	    facility
	  Planning
	  Scheduling
	  Routing
	  Inspection
	  Demonstration of a completely automated machine
	    shop consisting of computer-controlled NC machines
	    served by robots and robot carts, producing mixed
	    configurations of parts on demand

NASA
  $1 million/year
  12 person-years
  Langley Research Center,
    Hampton, VA
	Robotics laboratory to develop teleoperated and auto-
	    mated spacecraft servicing techniques
	Trajectory planning and multiarm coordination
	Human-machine interface for remote systems
	Operating system for concurrent processes in a dis-
	    tributed computer network
	Robotics simulation

Marshall Space Flight Center, Huntsville, AL	Research on end effectors Human-machine interface for teleoperation
JPL, Pasadena, CA	Research in hand-eye coordination Control of structurally flexible manipulators Teleoperators Three-fingered hand with Stanford Univ.

Naval Ocean Systems Center,
  San Diego, CA
  $1 million/year
	Exploring application of robot and teleoperator systems
	  Submersibles
	  Land vehicles
	  Underwater manipulators
	  Stereo optic and acoustic vision
	  Remote presence
	  Autonomous robots—knowledge representation and
	    automatic decision making
	  Complex robot systems specification and verification

Naval Ocean Systems Center,
  Hawaii
  $0.5 million/year
	Exploring the use of telepresence (teleoperation with
	    sensory feedback directly to the operator, giving a
	    sense of being at the work site) for use in materials
	    handling aboard ship

24-3. RESEARCH IN EUROPE AND THE USSR

The Western European countries are estimated to be spending two to four times as much as the United States is spending on robotics and related research. [A recent survey of teleoperator and robotics research in Europe indicates that $10 million per year is being spent in telemanipulation R&D alone (Plantier et al., 1981, p. 8).] As in the United States, the emphasis is on making robots more intelligent and versatile. "Germany, Italy, Sweden, France, Switzerland and the United Kingdom, either in national research laboratories or in robot manufacturers, are presently the most advanced European countries in computer vision and sensor-based robots" (Plantier et al., 1981, p. 12).

No authoritative figures are available on robotic research in the USSR and the Eastern block countries, but they appear to be making a strong effort in robotics in order to improve productivity and overcome labor shortages. However, the Soviets, lack of ready access to sophisticated digital and computer circuitry is hindering their research efforts in advanced robots (Teschler, 1981, p. 48).

25

APPLICATIONS NOW AND

IN THE FUTURE

Figure 25-1 indicates current and anticipated future use of robots in manufacturing in Japan. Currently, machine tool processing is dominant, followed by welding, assembly, and plastic molding. In terms of percentage of value, assembly, measurement, and inspection are expected to increase, while spot welding, plastic molding, and machining processes are expected to decrease.

Table 25-1 indicates 1981 and 1990 estimates of major uses for industrial robots in the United States. In general, the same trends are visible in the United States as in Japan, with welding and materials handling being currently dominant, moving toward assembly being the dominant application in 1990 (some estimates for assembly robots in 1990 ranging as high as 50 percent of all robots produced).

The results of a recent JIRA forecast (Hasegawa, 1981, p. 171) of robot applications development, introduction, and diffusion are summarized in Table 25-2. Observe that in the mid-1980s it is anticipated that robots for assembly, inspection, reactor maintenance, cleaning, and ocean development, as well as all phases of manufacturing, will be in commercial use, with diffusion relatively complete by 1990. In the last half of this decade it is forecast that construction and mining robots, robots for firefighting, and even robots to assemble soft goods such as textiles will begin commercial diffusion. In the same time period, prosthetic arms and legs, and robots for use in patient care and other medical applications, will emerge.

In the late 1980s, robots for use in various military applications, such as for use in dangerous missions and hostile environments and previously people-intensive activities such as weapons loading are also expected to appear. Also in the late 1980s, robots for use in servicing of space satellites are anticipated. In the 1990s,

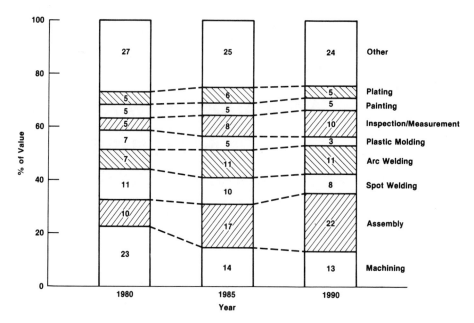

Figure 25-1 JIRA forecast of Japanese user needs for IRs by production process. (From Research Division of the Long-Term Credit Bank of Japan, 1981, p. 74.) Used by permission of JIRA and Survey Japan, Tokyo, Japan.

TABLE 25-1 Estimated U.S. Robot Order Trends by Usage

	Through 1981 (%)	1990 (%)
Materials handling, including machine loading and unloading	25–30	30–35
Assembly	10	35–40
Spotwelding	35–45	3–5
Arcwelding	5–8	15–20
Paint spraying	8–12	5
Other	8–10	7–10

Source: Conigliaro (1981, p. 6). Used by permission of *Robotics Age.*

robots for use in the construction and assembly of large space structures and in space manufacturing are forecast. Starting around 1995, semiautonomous robots are foreseen for use on extraterrestrial surfaces for exploration, mining, materials processing, manufacture, and the construction, tending, maintenance, and repair of space installations.

The general trend is toward robots with ever-increasing intelligence and sensory capabilities. Engelberger (1980, p. 135) suggests that robots will eventually be

TABLE 25-2   Future Commercial Applications of Robots

In commercial use in first half of 1980s, diffusion complete by 1990:
- All phases of manufacturing
- Assembly
- Inspection
- Reactor maintenance
- Cleaning
- Ocean development

Begin commercial diffusion:

(1) *In last half of decade:*
- Construction
- Firefighting
- Mining
- Soft goods assembly
- Patient care and other medical applications
- Prosthetics

(2) *In late 1980s*
- Military
  Dangerous missions
  Hostile environments
  People-intensive activities
- Space
  Servicing of space satellites
  Assembly and servicing of space-craft on earth
- Home: household maintenance
- Commercial
  Service industry application
  Equipment operation requiring intelligence

(3) *In the 1990s*
- Home: household repair
- Military
  Shipboard maintenance
  Equipment maintenance and repair
- Space
  Construction and assembly of large space structures
  Space manufacturing
- Commercial: equipment maintenance and repair

entering the service industries, with such diverse applications as garbage collection and fast-food preparation and dispensing. He even suggests that "a [limited] household robot may be practicable before the end of the 1980s decade."

26

THE ROBOT AND THE AUTOMATED
FACTORY

Most robots today are used in manufacturing as mechanical replacements for
formerly manual operations. However, robots are moving from being hardware-
intensive mechanical systems operating on a stand-alone basis to being highly intelligent software-intensive machines integrated into their environment.

The most significant recent changes in robot technology have been the result
of changing computer power. One view of a modern robot is as a machine tool that
can be treated as a terminal to a computer.

It is useful to consider several levels of manufacturing automation.

- *Stand-alone automation:* an individual robot; NC (numerically controlled)
  machine
- *Island of automation:* several computer-controlled machines (and robots)
  grouped together as a unit (a work cell) to perform a specific function
- *Flexible automation (FA):* a manufacturing process carried out by a collection of computer-controlled machines
- *Computer-integrated manufacturing (CIM):* an overall manufacturing system
  in which design, planning, management, job shop scheduling, and so on, are
  all under the coordinating control of a central computer

The following definitions and acronyms will also be used in the remaining
text.

- Work cell: a grouping of machines to do a specific function
- CAD: computer-aided design

- CAM: computer-aided manufacture
- CAE: computer-aided engineering

Ideally, a CIM enterprise would use CAE in conjunction with CAD to design the product. The resultant product data base would then be used to generate the process commands for CAM. The manufacturing system would then accept batch orders for various items and schedule their manufacture optimally in terms of priority and machine availability. Based on the scheduling, the raw stock would automatically be retrieved from the automated warehouse, inspected, and transported to the work cell for the first operation. The cell may have robots in attendance for part handling, inspection, and so on. After being processed by the first cell, using the retrieved process plan, the partially manufactured product would be inspected, passed on to other cells, and eventually sent to shipping or stored in an automated warehouse.

In 1983, there were probably fewer than a dozen CIMs that ever came close to such an operation. Urbaniak (1983) states: "Fanuc's new flexibly automated manufacturing plant at the base of Mt. Fuji is the closest thing to an unmanned factory in the world today." It employs approximately 60 people and 101 robots to service 60 machining cells and four assembly lines to produce 10,000 servomotors a month. All machinery and assembly operations are monitored by a central computer. It replaces the previous plant, where 108 people and 32 robots manufactured 6000 motors a month.

Most flexible automation operations are less robot intensive, with the computer-controlled machines being programmed to perform more handling operations internally than those performed by conventional machines—these computer-controlled machines often being serviced by transfer devices rather than robots.

It appears that Japan is not really ahead of the United States technically. However, many Japanese companies have emphasized manufacturing, implementing available technologies in efficient and useful ways. Dorf (1983) states that "though Japanese industry has surpassed the U.S. in the utilization of Programmable Automation and robots, such applications still account for only very limited sectors of their manufacturing industry and are even more scarcely used in Western Europe."

The United States leads in CAD and can enlarge on that to be leaders in CIM. The "factory automation" companies will generally be the largest corporations in the industry. These companies will sell robots as part of automating portions of a factory. Today, many computer factory systems exist as islands of automation. Tomorrow, we can expect that more integrated factory systems will become economically feasible. It appears that a major role of U.S. automation companies will be that of system integrators. Today, most manufacturing companies do not have the in-house expertise to perform the system design and integration themselves.

In addition to the many robotic newsletters referred to in Chapter 22, *Automation News*, available from Grant Publications, P.O. Box 1141, Dover, N.J. 07801, provides a means of keeping abreast of current activities in flexible manufacturing.

27

THE EXPANDING ROBOT POPULATION

Figure 27-1 projects the U.S. robotics industry growth through 1990, when it is estimated that the United States will have roughly 100,000 industrial robots (IRs) in operation–some 15 times that of 1982.

JIRA (Long-Term Credit Bank of Japan, 1981, p. 69) predicts that the demand for all classes of robots in Japan will grow at a 35% annual rate through 1985, and will grow at a 14% rate during the second half of the decade. Figure 27-2 provides JIRA's estimate of how this production will be distributed.

Integrating JIRA's forecast and the results of Table 14-1 and Fig. 27-1, Fig. 27-3 presents a composite forecast of the industrial robot population through 1990.* It appears that the robot population will pass 1,000,000 in the early 1990s and therefore will have a major impact on the world's productivity.

---

*Current trends indicate that this forecast is quite conservative, particularly with respect to the growth of the Japanese robot population.

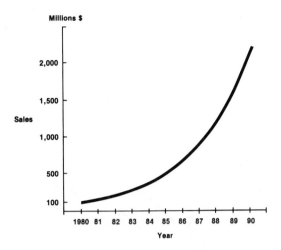

Figure 27-1 U.S. robotics industry growth 1980–1990. (Assumed average robot cost approximately $70,000 each.) Compound growth of revenue per year: 35%. (From Congilaro, 1981). Used by permission of *Robotics Age*.

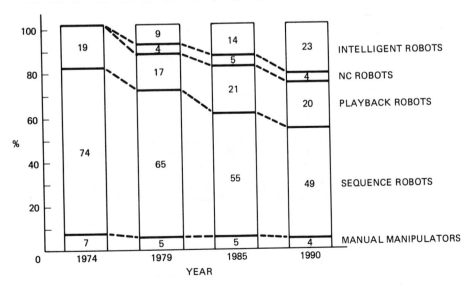

Figure 27-2 JIRA estimate of composition of IR production in Japan (based on value). Used by the permission of JIRA.

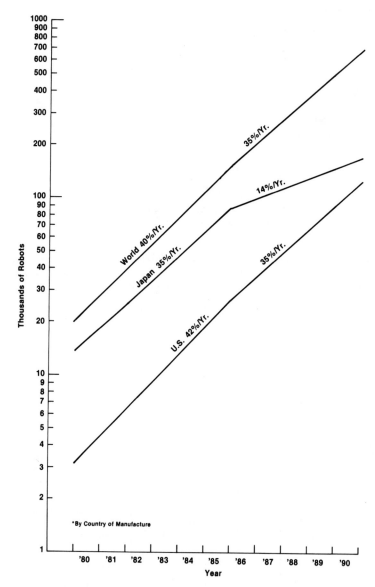

Figure 27-3 Forecast of industrial robot populations, 1980s decade, by country of manufacture. (Manual manipulators and fixed-sequence robots not included.)

28

ROBOTS AND THE SHAPE
OF THE FUTURE

Hasegawa (1981, pp. 169-170), observing the unfolding robot technology, forecasts that "industrial robots will become smarter, smaller, quicker, lighter, stronger, more ingenious, easier to operate, more intelligent and less expensive than they are now."

Engelberger (1980, p. 118) indicates that of the missing robot attributes in 1980, he considers the two crucial ones to be vision and tactile sensing, both of which are now entering use.

JIRA (Long-Term Credit Bank of Japan, 1981, p. 59) indicates that the technology for robot miniaturization, modularization, high-speed operation, and computer control has advanced to the point where it is now available and will grow in popularity and acceptance throughout this decade. Technology for part recognition and position determination is also currently available, but will not receive widespread popularity until the second half of this decade. Technology for color and behavior recognition, automated assembly, and machine tool and casting robotics will not become adequately advanced until after the middle of the decade and therefore will not achieve widespread popularity until the end of the decade.

One of the greatest advantages of robotics and automation occurs in cases where their use greatly increases equipment utilization. Albus (1981b) indicates that a robot arc welder cannot weld faster than a human being can, but by keeping its torch on the work about 90% of the time compared to no more than 30% of the time for a human being, it can turn out three times as much work in the same time. If through automation more shifts can be worked, productivity is increased even further. Unfortunately, present-day robots cannot set up their own work, so this reduces their productivity advantage. Albus conjectures:

Eventually, welding robots will be sufficiently sophisticated to work from plans stored in computer memory and to correct errors which may occur during a job. Welding robots will then be able to work nights and weekends (four shifts per week) completely without human supervision. At that point productivity improvements over present methods of many hundreds of percent become possible. Unfortunately, we are a long way from that today. There are many difficult research and development problems that must be solved first. Unless the level of effort in software development is increased many fold, these improvements will not be realized for many years.

In the metal-cutting industry, Albus (pp. 16-18) predicts:

During the 1980's, robot sensory and control capabilities will improve to the point where robots can find and load unoriented parts, or in some cases, even pick parts out of a bin filled with randomly oriented parts lying on top of each other. This may improve productivity by hundreds of percent because it will make it possible to install robots in many existing plants without major reengineering of production methods. . . .

By 1990, robots may begin to have a significant impact on mechanical assembly. There has been a great deal of research effort spent on robot assembly. Unfortunately, the results have not been spectacular—yet. On the one hand, robots cannot compete with classical so-called "hard automation" in assembly of mass produced parts. General purpose machines like robots are still too slow and too expensive to be economical for mass production assembly tasks. On the other hand, robots cannot yet compete with human assembly workers in small lot assembly. Humans are incredibly adaptable, dexterous, as well as fast, skilled, and relatively cheap compared to robots. . . .

Nevertheless, progress is being made and will continue. Robot capabilities will gradually increase. Sensory systems will become more sophisticated and less expensive. The cost of computer hardware is dropping rapidly and steadily with no sign of bottoming out. Software costs are likely to be the major impediment to robot development of the foreseeable future, but even these are slowly yielding to the techniques of structured programming and high level languages.

Eventually, extremely fast, accurate, dexterous robots will be programmed using design graphics data bases which describe the shape of the parts to be made and the configuration of the assemblies to be constructed. Eventually, robots will be able to respond to a wide variety of sensory cues, to learn by experience and to acquire skills by self optimization. Such skills can then be transferred to other robots so that learning can be propagated rapidly throughout the robot labor force.

Sometime, perhaps around the turn of the century, robot technology will develop to the degree necessary to produce the totally automated factory. In such factories, robots will perform most, if not all, of the operations that now require human skills. There will be totally automatic inventory and tool management, automatic machining, assembly, finishing, and inspection systems. Automatic factories will even be able to reproduce themselves. That is, automatic factories will make the components for other automatic factories.

Once this occurs, productivity improvements will propagate from generation to generation. Each generation of machines will produce machines less expensive and more sophisticated than themselves. This will bring about an exponential decline in the cost of robots and automatic factories which may equal the cost/performance record of the computer industry. . . .

Eventually, products produced in automatic factories may cost only slightly more than the raw materials and energy from which they are made.

In summary, what we see emerging are robots with increasing intelligence, sensory capability, and dexterity. Initially, we will see an increasing use of off-line programming of computer-controlled robots, using improved robot command languages. Provision will be made to include the role of sensors, such as vision and touch, in this programming. Later, self-planning will emerge as higher and more general commands are given to the robot. At this point, the marriage of robotics and artificial intelligence will be virtually complete. At the same time as all this is taking place, robotic hands with improved dexterity (so that it will not be necessary to move the entire arm for fine motions), and advanced control systems to support this dexterity, will emerge. Also emerging will be robots with coordinated multiple arms and eventually even legs, supported by even more sophisticated control systems.

As this evolution progresses, information and intelligence will become the dominant factors in robotics, with the manipulator devices and sensors shrinking in importance to the skeleton that undergirds this dominating "ghost in the machine."

REFERENCES FOR PART II

Albus, J. S., *Brains, Behavior, and Robotics.* Peterborough, NH: Byte Books, 1981a.

Albus, J. S., "Industrial Robots and Productivity Improvement," *Summary and Report of an Exploratory Workshop on the Social Impact of Robotics,* Office of Technology Assessment, Washington, DC, July 24, 1981b.

Allen, J. F., and **Koomen, J. A.,** "Planning Using a Temporal World Model," *Proceedings of the Eighth International Joint Conference on AI,* Karlsruhe, West Germany, Aug. 8-12, 1983. Los Altos, CA: W. Kaufmann, 1983, pp. 741-747.

Aron, P., "The Robot Scene in Japan: An Update," Paul Aron Report 26, Daiwa Securities of America, Inc., New York, Sept. 7, 1983.

Barrow, H. G., and **Tenenbaum, J. M.,** "Computational Vision," *Proceedings of the IEEE,* Vol. 69, No. 5, May 1981, pp. 575-579.

Birk, J. R., and **Kelley, R. B.,** "An Overview of the Basic Research Needed to Advance the State of Knowledge in Robotics," *IEEE Transactions on Systems, Man and Cybernetics,* Vol. SMC-11, No. 8, Aug. 1981, pp. 575-579.

Congliaro, L., "Robotics Presentation, Institutional Investors Conference: May 28, 1981," *Bache Robotics Newsletter* 81-429, Bache Halsey Stuart Shields, Inc., New York, June 19, 1981.

Congliaro, L., "Trends in the Robot Industry (Revisited): Where Are We Now?" *Proceedings of the 13th International Symposium on Industrial Robots and Robot 7,* Chicago, Apr. 17-21, 1983, pp. 1-1-1-17.

Dorf, R. C., "Robotics and the Automated Manufacturing Industry," *Proceedings of the 13th International Symposium on Industrial Robots and Robots 7,* Chicago, Apr. 17-21, 1983, pp. 1-25-1-40.

Electric Power Research Institute, *The Impact of Industrial Robots on Electric Loads in U.S. Industry*, EPRI EM-3325, Palo Alto, CA, Feb. 1984.

Engelberger, J. F., *Robots in Practice*. New York: American Management, 1980.

Gleason, G. J., and Agin, G. J., "A Modular Vision System for Sensor-Controlled Manipulation and Inspection," *Proceedings of the 9th International Symposium on Industrial Robots*, Washington, DC, Mar. 1979.

Gruver, W. A., Soroka, B. I., Craig, J. J., and Turner, T. L., "Evaluation of Commercially Available Robot Programming Languages," *Proceedings of the 13th International Symposium on Industrial Robots and Robots 7*, Chicago, Apr. 17–21, 1983, pp. 12-58–12-68.

Hasegawa, Y., "The Future of Industrial Robots," in *Robots in the Japanese Economy*, K. Sadamoto (Ed.). Tokyo: Survey Japan, 1981, pp. 151–174.

Kelly, F., and Huston, R., *Recent Advances in Robotics Research*, SAE Technical Paper 800383, Feb. 1980.

Kinnucan, P., "How Smart Robots are Becoming Smarter," *High Technology*, Vol. 1, No. 1, Sept./Oct. 1981, pp. 32–40.

Long Term Credit Bank of Japan, Research Division, "Industrial Robots in Japan," in *Robots in the Japanese Economy*, K. Sadamoto (Ed.). Tokyo: Survey Japan, 1981, pp. 1–77.

Lozano-Pérez, T., "Robot Programming," *Proceedings of the IEEE*, Vol. 71, No. 7, July 1983, pp. 821–841.

Malik, J., and Binford, T. O., "Reasoning in Time and Space," *Proceedings of the Eighth International Joint Conference on Artificial Intelligence*, Karlsruhe, West Germany, Aug. 8–12, 1983. Los Altos, CA: W. Kaufmann, 1983, pp. 343–345.

Marr, D., and Nishihara, H., "Visual Information Processing: Artificial Intelligence and the Sensorium of Sight," *Technology Review*, Oct. 1978, pp. 28–47.

Plantier, M., et al., "Teleoperation and Automation: A Survey of European Expertise Applicable to Docking and Assembly in Space," ESTEC Contract 4402/80/ NL/AK (SC), EUROSTAT, S.A., Geneva, May 1981.

Proceedings of NBS/Air Force ICAM Workshop on Robot Interfaces, NBSIR 80-2152, Jan. 1981.

Rheinhold, A. G., and Vanderbrug, G., "Robot Vision for Industry: The Autovision System," *Robotics Age*, Vol. 2, No. 3, Fall 1980, pp. 22–28.

"Robots Join the Labor Force," *Business Week*, June 9, 1980, pp. 62–76.

RIA Robotics Glossary, Dearborn, MI: Robot Institute of America, 1984.

Rosen, C. A., "Machine Vision and Robotics: Industrial Requirements," TN 174, SRI International, Menlo Park, CA, Nov. 1978.

Rossol, L., "Vision and Adaptive Robots in General Motors," *First International Conference on Robot Vision and Sensory Controls*, Stratford-upon-Avon, Apr. 13, 1981, pp. 277–287.

Rossol, L., "Computer Vision in Industry—The Next Decade," GMR-4096, GM Research Lab., Computer Science Dept., Warren, MI, July 22, 1982.

"Russian Robots Run to Catch Up," *Business Week*, Aug. 17, 1981, p. 120.

Saveriano, J. W., "Industrial Robots Today and Tomorrow," *Robotics Age*, Vol. 2, No. 2, Summer 1980, pp. 4–15.

Scheinman, V., "The Autovision System Computerized Vision for Factory Automation," *Proceedings of the 11th International Symposium on Industrial Robots*, Tokyo, Oct. 1981.

Simons, G. L., *Robots in Industry*. Manchester, England: The National Computer Center Ltd., 1980.

Tanner, W. R., "Basics of Robots," in *Industrial Robots, Vol. I: Fundamentals*, 2nd ed. W. R. Tanner (Ed.). Dearborn, MI: SME, 1981, pp. 3–12.

Teschler, L., "How Good Is Soviet Robot Technology?" *Machine Design*, Oct. 8, 1981, pp. 43–48.

Toepperwein, R., et al., "ICAM Robotics Application Guide (RAG)," ARWAL-TR-80-4042, Vol. II, Air Force Wright Aerospace Labs, Wright Patterson AFB, OH, Apr. 1980.

Umentani, Y., "Research and Development of IRs in Japan," in *Robots in the Japanese Economy*, K. Sadamato (Ed.). Tokyo: Survey Japan, 1981, pp. 70–104.

Urbaniak, D. F., "The Unattended Factory: FANUC's New Flexible Automatic Plant Using Industrial Robots," *Proceedings of the 13th International Symposium on Industrial Robots and Robots 7*, Chicago, Apr. 17–21, 1983, pp. 1-18-1-24.

Worldwide Robotics Survey and Directory, Dearborn, MI: Robot Institute of America, 1983.

Yonemoto, K., "The Art of Industrial Robots in Japan," *Proceedings of the 11th International Symposium on Industrial Robots*, Tokyo, Oct. 1981, pp. 1–7.

APPENDICES

The following six appendices are for use with Part I of the text.

A

SOURCES FOR
FURTHER INFORMATION ON AI

1. Journals
   - *SIGART Newsletter*—ACM (Association for Computing Machinery).
   - *Artificial Intelligence.*
   - *Cognitive Science*—Cognitive Science Society.
   - *AI Magazine*—American Association for AI (AAAI).
   - *Pattern Analysis and Machine Intelligence*—IEEE.
   - *International Journal on Robotics Research.*
   - *IEEE Transactions on Systems, Man and Cybernetics.*
   - *Expert Systems*
   - *Journal of Automated Reasoning*
   - *Future Generations Computer Systems*
   - *Speech Communication*
2. Conferences
   - *International Joint Conference on AI (IJCAI),* biannual (odd numbered years.)
   - AAAI Annual Conference.
   - IEEE Systems, Man and Cybernetics Annual Conference.
3. Recent Books
   - Barr, A., and Feigenbaum, E. A., *The Handbook of Artificial Intelligence*, Vols. I and II. Los Altos, CA: W. Kaufmann, 1981, 1982.
   - Bledsoe, W. W., and Loveland, D. W., *Automated Theorem Proving: After 25 Years*. Providence, RI: American Mathematical Society, 1984.

- Buchanan, B.G., and Shortliffe, E.H., *Rule-Based Expert Systems*. Reading, MA: Addison-Wesley, 1984.

- Clancey, W.J., and Shortliffe, E.H. (Eds.), *Readings in Medical Artificial Intelligence*. Reading, MA.: Addison-Wesley, 1984.

- Clocksin, W. F., and Mellish, C. S., *Programming in PROLOG*. New York: Springer-Verlag, 1981.

- Cohen, P. R., and Feigenbaum, E. A., (Eds.), *The Handbook of Artificial Intelligence*, Vol. III. Los Altos, CA: W. Kaufmann, 1982.

- Davis, R., and Lenat, D. B., *Knowledge-Based Systems in Artificial Intelligence*. New York: McGraw-Hill, 1982.

- Feigenbaum, E. A., and McCorduck, P. *The Fifth Generation*. Reading, MA: Addison-Wesley, 1983.

- Hayes-Roth, F. (Ed.), *Building Expert Systems*. Reading, MA: Addison-Wesley, 1983.

- Michalski, R. S., Carbonell, J. G., and Mitchell, T. M. (Eds.), *Machine Learning—An Artificial Intelligence Approach*. Palo Alto, CA: Tioga, 1983.

- Nilsson, N. J., *Principles of Artificial Intelligence*. Palo Alto, CA: Tioga, 1980.

- Rich, E., *Artificial Intelligence*. New York: McGraw-Hill, 1983.

- Simon, H. A., *The Sciences of the Artificial*, 2nd ed. Cambridge, MA: MIT Press, 1981.

- Shank, R. C., with Childers, P.G., *The Cognitive Computer*. Reading, MA.: Addison-Wesley, 1984.

- Sowa, J. F., *Conceptual Structures, Information Processing in Mind and Machine*. Reading, MA: Addison-Wesley, 1983.

- Szolovits, P. (Ed.), *Artificial Intelligence in Medicine*. Boulder, CO: Westview Press, 1982.

- Weiss, S. M., and Kulikowski, C. A., *A Practical Guide to Designing Expert Systems*. Totowa, NJ: Rowman & Allanheld, 1984.

- Wilensky, R. *Planning and Understanding*. Reading, MA: Addison-Wesley, 1982.

- Wilensky, R., *LISPcraft*. New York: W.W. Norton, 1984.

- Winston, P. H., *Artificial Intelligence*, 2nd ed. Reading, MA: Addison-Wesley, 1984.

- Winston, P. H., and Horn, B. K. P., *LISP.*, 2nd ed. Reading, MA: Addison-Wesley, 1984.

- Winston, P. H., and Prendergast, K. A. (Eds.), *The AI Business*. Cambridge, MA.: MIT Press, 1984.

- Wos, L., Overbeek, R., Lusk, E., and Boyle, J., *Automated Reasoning: Introduction and Applications*. Englewood Cliffs, NJ: Prentice-Hall, 1984.

4. Newsletters
   - *The Artificial Intelligence Report*, Artificial Intelligence Publications, 3600 W. Bayshore Rd., Palo Alto, CA 94303.
   - *Applied Artificial Intelligence Reporter*, Intelligent Computer Systems Research Institute, University of Miami, P.O. Box 1308-EP, Fort Lee, NJ 07024.
5. NASA/NBS Series of Overview Reports on Artificial Intelligence and Robotics. Available from the National Technical Information Service (NTIS), Springfield, VA 22161. (The asterisked reports are also available in a reformatted form in a library binding as *Robotics and Artificial Intelligence Applications*, from Business/Technology Books, P.O. Box 574, Orinda, CA 94563.)

   *An Overview of Expert Systems, NBSIR 2505, May 1982 (rev. Oct. 1982).

   *An Overview of Computer Vision, NBSIR 2582, Sept. 1982.

   *An Overview of Natural Language Processing, NBSIR 83-2687, Apr. 1983, NASA TM 85635, Apr. 1983.

   An Overview of Artificial Intelligence and Robotics, Vol. I: *Artificial Intelligence.*

   Part A: *The Core Ingredients*, NASA TM 85836, June 1983, NBSIR 83-2799. Jan. 1984.

   Part B: *Applications*, NASA TM 85838, Oct. 1983.

   Part C: *Basic AI Topics*, NASA TM 85839, Dec. 1983.

   The important AI application areas of robotics and automated manufacturing are treated in:

   *An Overview of Artificial Intelligence and Robotics, Vol. II: *Robotics*, NBSIR 82-2479, Mar. 1982.

B

GLOSSARY OF AI TERMS

A

Activation Mechanism: The situation required to invoke a procedure—usually a match of the system state to the preconditions required to exercise a production rule.

Algorithm: A procedure for solving a problem in a finite number of steps.

AND/OR Graph: A generalized representation for problem reduction situations and two-person games. A treelike structure with two types of nodes. Those nodes which all have to be accomplished (or considered) are AND nodes. Those nodes for which only one of several is necessary are OR nodes. (In about half the literature the labeling of AND and OR nodes is reversed from this definition.)

Antecedent: The left-hand side of a production rule. The pattern needed to make the rule applicable.

Argument Form: A reasoning procedure in logic.

ARPANET: A network of computers and computational resources used by the U.S. AI community and sponsored by DARPA (Defense Advanced Research Projects Agency).

Artificial Intelligence (AI): A discipline devoted to developing and applying computational approaches to intelligent behavior. Also referred to as machine intelligence or heuristic programming.

Artificial Intelligence Approach: An approach that has as its emphasis symbolic processes for representing and manipulating knowledge in a problem-solving mode.

Atom: An individual. A proposition in logic that cannot be broken down into other propositions. An indivisible element.

Autonomous: A system capable of independent action.

B

Backtracking: Returning (usually due to depth-first search failure) to an earlier point in a search space. Also a name given to depth-first backward reasoning.

Backward Chaining: A form of reasoning starting with a goal and recursively chaining backward to its antecedent goals or states by applying applicable operators until an appropriate earlier state is reached or the system backtracks. This is a form of depth-first search. When the application of operators changes a single goal or state into multiple goals or states, the approach is referred to as a problem reduction.

Blackboard Approach: A problem-solving approach whereby the various system elements communicate with each other via a common working data storage called a blackboard.

Blind Search: An ordered approach that does not rely on knowledge for searching for a solution.

Blocks World: A small artificial world, consisting of blocks and pyramids, used to develop ideas in computer vision, robotics, and natural language interfaces.

Bottom-Up Control Structure: A problem-solving approach that employs forward reasoning from current or initial conditions. Also referred to as an event-driven or data-driven control structure.

Breadth-First Search: An approach in which, starting with the root node, the nodes in the search tree are generated and examined level by level (before proceeding deeper). This approach is guaranteed to find an optimal solution if it exists.

C

Clause: A syntactic construction containing a subject and a predicate and forming part of a statement in logic or part of a sentence in a grammar.

Cognition: An intellectual process by which knowledge is gained about perceptions or ideas.

Combinatorial Explosion: The rapid growth of possibilities as the search space expands. If each branch point (decision point) has an average of n branches, the search space tends to expand as n^d as the depth of search, d, increases.

Common Sense: The ability to act appropriately in everyday situations based on one's lifetime accumulation of experiential knowledge.

Commonsense Reasoning: Low-level reasoning based on a wealth of experience.

Compile: The act of translating a computer program written in a high-level language (such as LISP) into the machine language which controls the basic operations of the computer.

Computational Logic: A science designed to make use of computers in logic calculus.

Computer Architecture: The manner in which various computational elements are interconnected to achieve a computational function.

Computer Graphics: Visual representations generated by a computer (usually observed on a monitoring screen).

Computer Network: An interconnected set of communicating computers.

Computer Vision (Computational or Machine Vision): Perception by a computer, based on visual sensory input, in which a symbolic description is developed of a scene depicted in an image. It is often a knowledge-based, expectation-guided process that uses models to interpret sensory data. Used somewhat synonymously with image understanding and scene analysis.

Conceptual Dependency: An approach to natural language understanding in which sentences are translated into basic concepts expressed as a small set of semantic primitives.

Conflict Resolution: Selecting a procedure or rule from a conflict set of applicable competing procedures or rules.

Conflict Set: The set of rules that matches some data or pattern in the global data base.

Conjunct: One of several subproblems. Each of the component formulas in a logical conjunction.

Conjunction: A problem composed of several subproblems. A logical formula built by connecting other formulas by logical ANDs.

Connectives: Operators (e.g., AND, OR) connecting statements in logic so that the truth value of the composite is determined by the truth value of the components.

Consequent: The right side of a production rule. The result of applying a procedure.

Constraint Propagation: A method for limiting search by requiring that certain constraints be satisfied. It can also be viewed as a mechanism for moving information between subproblems.

Context: The set of circumstances or facts that define a particular situation, event, and so on. The portion of the situation that remains the same when an operator is applied in a problem-solving situation.

Control Structure: Reasoning strategy. The strategy for manipulating the domain knowledge to arrive at a problem solution.

D

Data Base: An organized collection of data about some subject.

Data Base Management System: A computer system for the storage and retrieval of information about some domain.

Data-Driven: A forward-reasoning, bottom-up problem-solving approach.

Data Structure: The form in which data are stored in a computer.

Debugging: Correcting errors in a plan.

Declarative Knowledge Representation: Representation of facts and assertions.

Deduction: A process of reasoning in which the conclusion follows from the premises given.

Default Value: A value to be used when the actual value is unknown.

Depth-First Search: A search that proceeds from the root node to one of the successor nodes and then to one of that node's successor nodes, and so on, until a solution is reached or the search is forced to backtrack.

Difference Reduction: "Means-ends" analysis. An approach to problem solving that tries to solve a problem by iteratively applying operators that will reduce the difference between the current state and the goal state.

Directed Graph: A knowledge representation structure consisting of nodes (representing, e.g., objects) and directed connecting arcs (labeled edges, representing, e.g., relations).

Disproving: An attempt to prove the impossibility of a hypothesized conclusion (theorem) or goal.

Domain: The problem area of interest: for example, bacterial infections, prospecting, VLSI design.

E

Editor: A software tool to aid in modifying a software program.

Embed: To write a computer language on top of (embedded in) another computer language (such as LISP).

Emulate: To perform like another system.

Equivalent: Has the same truth value (in logic).

Evaluation Function: A function (usually heuristic) used to evaluate the merit of the various paths emanating from a node in a search tree.

Event-Driven: A forward-chaining problem-solving approach based on the current problem status.

Expectation-Driven: Processing approaches that proceed by trying to confirm models, situations, states, or concepts anticipated by the system.

Expert System: A computer program that uses knowledge and reasoning techniques to solve problems normally requiring the abilities of human experts.

F

Fault Diagnosis: Determining the trouble source in an electromechanical system.

Fifth-Generation Computer: A non-von Neumann, intelligent, parallel-processing form of computer now being pursued by Japan.

First-Order Predicate Logic: A popular form of logic used by the AI community for representing knowledge and performing logical inference. First-order predicate logic permits assertions to be made about variables in a proposition.

Forward Chaining: Event-driven or data-driven reasoning.

Frame: A data structure for representing stereotyped objects or situations. A frame has slots to be filled for objects and relations appropriate to the situation.

FRANZLISP: The dialect of LISP developed at the University of California, Berkeley.

Functional Application: The generic task or function performed in an application.

Fuzzy Set: A generalization of set theory that allows for various degrees of set membership, rather than all or none.

G

Garbage Collection: A technique for recycling computer memory cells no longer in use.

General Problem Solver (GPS): The first problem solver (1957) to separate its problem-solving methods from knowledge of the specific task being considered. The GPS problem-solving approach employed was "means-ends analysis."

Generate and Test: A common form of state-space search based on reasoning by elimination. The system generates possible solutions and the tester prunes those solutions that fail to meet appropriate criteria.

Global Data Base: Complete data base describing the specific problem, its status, and that of the solution process.

Goal-Driven: A problem-solving approach that works backward from the goal.

Goal Regression: A technique for constructing a plan by solving one conjunctive subgoal at a time, checking to see that each solution does not interfere with the other subgoals that have already been achieved. If interferences occur, the offending subgoal is moved to an earlier noninterfering point in the sequence of subgoal accomplishments.

Graph: A set of nodes connected by arcs.

H

Heuristic Search Techniques: Graph searching methods that use heuristic knowledge about the domain to help focus the search. They operate by generating and testing intermediate states along potential solution paths.

Heuristics: Rules of thumb or empirical knowledge used to help guide a problem solution.

Hierarchical Planning: A planning approach in which first a high-level plan is formulated considering only the important (or major) aspects. Then the major steps of the plan are refined into more detailed subplans.

Hierarchy: A system of things ranked one above the other.

Higher-Order Language (HOL): A computer language (such as FORTRAN or LISP) requiring fewer statements than machine language and usually substantially easier to use and read.

Horn Clause: A set of statements joined by logical ANDS. It has at most only one conclusion. Used in PROLOG.

I

Identity: Two propositions (in logic) that have the same truth value.

Image Understanding (IU): Visual perception by a computer employing geometric modeling and the AI techniques of knowledge representation and cognitive processing to develop scene interpretations from image data. IU has dealt extensively with three-dimensional objects.

Implies: A connective in logic that indicates that if the first statement is true, the statement following is also true.

Individual: A nonvariable element (or atom) in logic that cannot be broken down further.

Infer: To derive by reasoning. To conclude or judge from premises or evidence.

Inference: The process of reaching a conclusion based on an initial set of propositions, the truths of which are known or assumed.

Inference Engine: Another name given to the control structure of an AI problem solver in which the control is separate from the knowledge.

Instantiation: Replacing a variable by an instance (an individual) that satisfies the system (or satisfies the statement in which the variable appears).

Intelligence: The degree to which an individual can successfully respond to new situations or problems. It is based on the individual's knowledge level and the ability to appropriately manipulate and reformulate that knowledge (and incoming data) as required by the situation or problem.

Intelligent Assistant: An AI computer program (usually an expert system) that aids a person in the performance of a task.

Interactive Environment: A computational system in which the user interacts (dialogues) with the system (in real time) during the process of developing or running a computer program.

Interface: The system by which the user interacts with the computer. In general, the junction between two components.

INTERLISP: A dialect of LISP (used at Stanford University) developed at Bolt, Beranek, and Newman and Xerox PARC.

Invoke: To place into action (usually by satisfying a precondition).

K

Knowledge Base: AI data bases that are not merely files of uniform content, but collections of facts, inferences, and procedures corresponding to the types of information needed for problem solution.

Knowledge Base Management: Management of a knowledge base in terms of storing, accessing, and reasoning with the knowledge.

Knowledge Engineering: The AI approach focusing on the use of knowledge (e.g., as in expert systems) to solve problems.

Knowledge Representation (KR): The form of the data structure used to organize the knowledge required for a problem.

Knowledge Source: An expert system component that deals with a specific area or activity.

L

Leaf: A terminal node in a tree representation.

Least Commitment: A technique for coordinating decision making with the availability of information, so that problem-solving decisions are not made arbitrarily or prematurely, but are postponed until there is enough information.

List: A sequence of zero or more elements enclosed in a pair of parentheses, where each element is either an atom (an indivisible element) or a list.

List Processing Language (LISP): The basic AI programming language.

Literal: An atom (e.g., q) or an atom preceded by NOT (e.g., NOT q).

Logical Operation: Execution of a single computer instruction.

Logical Representation: Knowledge representation by a collection of logical formulas (usually in first-order predicate logic) that provide a partial descrption of the world.

M

MACLISP: A dialect of LISP developed at MIT.

Means-Ends Analysis: A problem-solving approach (used by GPS) in which problem-solving operators are chosen in an iterative fashion to reduce the difference between the current problem-solving state and the goal state.

Meta-Rule: A higher-level rule used to reason about lower-level rules.

Microcode: A computer program at the basic machine level.

Model Driven: A top-down approach to problem solving in which the inferences to be verified are based on the domain model used by the problem solver.

Modus Ponens: A mathematical form of argument in deductive logic. It has the form

If A is true, then B is true.

A is true.

Therefore, B is true.

N

Natural Deduction: Informal reasoning.

Natural Language Interface (NLI): A system for communicating with a computer by using a natural language.

Natural Language Processing (NLP): Processing of natural language (e.g., English) by a computer to facilitate communication with the computer or for other purposes, such as language translation.

Natural Language Understanding (NLU): Response by a computer based on the meaning of a natural language input.

Negate: To change a proposition into its opposite.

Node: A point (representing such aspects as the system state or an object) in a graph connected to other points in the graph by arcs (usually representing relations).

Nonmonotonic Logic: A logic in which results are subject to revision as more information is gathered.

O

Object-Oriented Programming: A programming approach focused on objects that communicate by message passing. An object is considered to be a package of information and descriptions of procedures that can manipulate that information.

Operators: Procedures or generalized actions that can be used for changing situations.

P

Parallel Processing: Simultaneous processing, as opposed to the sequential processing in a conventional (von Neumann) type of computer architecture.

Path: A particular track through a state graph.

Pattern Directed Invocation: The activation of procedures by matching their antecedent parts to patterns present in the global data base (the system status).

Pattern Matching: Matching patterns in a statement or image against patterns in a global data base, templates, or models.

Pattern Recognition: The process of classifying data into predetermined categories.

Perception: An active process in which hypotheses are formed about the nature of the environment, or sensory information is sought to confirm or refute hypotheses.

Personal AI Computer: New, small, interactive, stand-alone computers for use by an AI researcher in developing AI programs. Usually specifically designed to run an AI language such as LISP.

Plan: A sequence of actions to transform an initial situation into a situation satisfying the goal conditions.

Portability: The ease with which a computer program developed in one programming environment can be transferred to another.

Predicate: That part of a proposition that makes an assertion (e.g., states a relation or attribute) about individuals.

Predicate Logic: A modification of propositional logic to allow the use of variables and functions of variables.

Prefix Notation: A list representation (used in LISP programming) in which the connective, function, symbol, predicate is given before the arguments.

Premise: A first proposition on which subsequent reasoning rests.

Problem Reduction: A problem-solving approach in which operators are used to change a single problem into several subproblems (which are usually easier to solve).

Problem Solving: A procedure using a control strategy to apply operators to a situation to try to achieve a goal.

Problem State: The condition of the problem at a particular instant.

Procedural Knowledge Representation: A representation of knowledge about the world by a set of procedures—small programs that know how to do specific things (how to proceed in well-specified situations).

Production Rule: A modular knowledge structure representing a single chunk of knowledge, usually in if-then or antecedent-consequent form. Popular in expert systems.

Programming Environment: The total programming setup, including the interface, the languages, the editors, and other programming tools.

Programming in Logic (PROLOG): A logic-oriented AI language developed in France and popular in Europe and Japan.

Property List: A knowledge representation technique by which the state of the world is described by objects in the world via lists of their pertinent properties and their associated attributes and values.

Proposition: A statement (in logic) that can be true or false.

Propositional Logic: An elementary logic that uses argument forms to deduce the truth or falsehood of a new proposition from known propositions.

Prototype: An initial model or system that is used as a base for constructing future models or systems.

Pseudoreduction: An approach to solving the difficult problem case where multiple goals must be satisfied simultaneously. Plans are found to achieve each goal independently and then integrated using knowledge of how plan segments can be intertwined without destroying their important effects.

R

Recursive Operations: Operations defined in terms of themselves.

Relaxation Approach: An iterative problem-solving approach in which initial conditions are propagated utilizing constraints until all goal conditions are adequately satisfied.

Relevant Backtracking (Dependency-Directed or Nonchronological Backtracking): Backtracking (during a search) not to the most recent choice point, but to the most relevant choice point.

Resolution: A general, automatic, syntactic method for determining if a hypothesized conclusion (theorem) follows from a given set of premises (axioms).

Root Node: The initial (apex) node in a tree representation.

Rule Interpreter: The control structure for a production rule system.

S

Satisficing: Developing a satisfactory, but not necessarily optimum solution.

Scheduling: Developing a time sequence of things to be done.

Scripts: Framelike structures for representing sequences of events.

Search Space: The implicit graph representing all the possible states of the system which may have to be searched to find a solution. In many cases the search space is infinite. The term is also used for non-state-space representations.

Semantic: Of or relating to meaning.

Semantic Network: A knowledge representation for describing the properties and relations of objects, events, concepts, situations, or actions by a directed graph consisting of nodes and labeled edges (arcs connecting nodes).

Semantic Primitives: Basic conceptual units in which concepts, ideas, or events can be represented.

S-Expression: A symbolic expression. In LISP, a sequence of zero or more atoms or S-expressions enclosed in parentheses.

Slot: An element in a frame representation to be filled with designated information about the particular situation.

Software: A computer program.

Solution Path: A successful path through a search space.

Speech Recognition: Recognition by a computer (primarily by pattern matching) of spoken words or sentences.

Speech Synthesis: Developing spoken speech from text or other representations.

Speech Understanding: Speech perception by a computer.

SRI Vision Module: An important object recognition, inspection, orientation, and location research vision system developed at SRI. This system converted the scene into a binary image and extracted the calculated needed vision parameters in real time, as it sequentially scanned the image line by line.

State Graph: A graph in which the nodes represent the system state and the connecting arcs represent the operators which can be used to transform the state from which the arcs emanate to the state at which they arrive.

Stereotyped Situation: A generic, recurrent situation such as "eating at a restaurant" or "driving to work."

Subgoals: Goals that must be achieved to achieve the original goal.

Subplan: A plan to solve a portion of the problem.

Subproblems: The set of secondary problems that must be solved to solve the original problem.

Syllogism: A deductive argument in logic whose conclusion is supported by two premises.

Symbolic: Relating to the substitution of abstract representations (symbols) for concrete objects.

Syntax: The order of arrangement (e.g., the grammar) of a language.

T

Terminal Node (Leaf Node): The final node emanating from a branch in a tree or graph representation.

Theorem: A proposition, or statement, to be proved based on a given set of premises.

Theorem Proving: A problem-solving approach in which a hypothesized conclusion (theorem) is validated using deductive logic.

Time Sharing: A computer environment in which multiple users can use the computer virtually simultaneously via a program that time-allocates the use of computer resources among the users in a near-optimum manner.

Top-Down Approach: An approach to problem solving that is goal-directed or expectation-guided based on models or other knowledge. Sometimes referred to as "hypothesize and test."

Top-Down Logic: A problem-solving approach used in production systems, where production rules are employed to find a solution path by chaining backward from the goal.

Tree Structure: A graph in which one node, the root, has no predecessor node, and all other nodes have exactly one predecessor. For a state-space representation, the tree starts with a root node (representing the initial problem situation). Each of the new states that can be produced from this initial state by application of a single operator is represented by a successor node of the root node. Each successor node branches in a similar way until no further states can be generated or a solution is reached. Operators are represented by the directed arcs from the nodes to their successor nodes.

Truth Maintenance: A method of keeping track of beliefs (and their justifications) developed during problem solving, so that if contradictions occur, the incorrect beliefs or lines of reasoning, and all conclusions resulting from them, can be retracted.

Truth Value: One of the two possible values—true or false—associated with a proposition in logic.

U

Unification: The name for the procedure for carrying out instantiations. In unification, the attempt is to find substitutions for variables that will make two atoms identical.

V

Variable: A quantity or function that may assume any given value or set of values.
Von Neumann Architecture: The current standard computer architecture that uses
  sequential processing.

W

World Knowledge: Knowledge about the world (or domain of interest).
World Model: A representation of the current situation.

C

AI LANGUAGES, TOOLS, AND
COMPUTERS

C-1. PROGRAMMING NEEDS OF AI

AI research has been an experimental science trying to develop computer programs that exhibit intelligent behavior. This has proven to be a difficult endeavor requiring the best programming tools. AI programs tend to develop iteratively and incrementally. As the programs are thus evolutionary, creating AI programs requires an interactive environment with built-in aids such as dynamic allocation of computer memory as the program evolves, rather than advance memory allocation as in most other programming domains. More important, the unpredictable intermediate forms of the data (as the program evolves) also influence the form of the programming languages and the management of memory.

Another aspect of AI programming is that AI researchers found that a great simplification in writing programs could be achieved by expressing functions recursively (defined in terms of themselves). Thus AI programming languages tend to support recursive processing. Finally, AI programs are concerned primarily with symbol manipulation rather than numeric computation. All AI languages thus support this feature.

Barr and Feingenbaum (1982, p. 32) observe that "AI programs are among the largest and most complex computer programs ever developed and present formidable design and implementation problems. . . . AI researchers in their capacity as language designers and programmers have pioneered an interactive mode of programming in environments with extensive support: editors, trace and debugging packages, and other aids for the construction of large complex systems."

Two basic general AI languages—LISP and PROLOG—have evolved in answer to these programming requirements. LISP has been the primary AI programming language. PROLOG, a logic-based language, has appeared more recently and has gained favor in Europe and Japan.

Various derivatives and dialects of LISP exist. Special high-level programming languages, for such purposes as assisting in knowledge representation and constructing expert systems, have been built on top of LISP and PROLOG.

In recent years, nearly all AI programs were developed on the DEC PDP-10 and PDP-11 computers. AI programming is now transitioning to the DEC VAX computers and the new personal AI machines.

C-2. LIST REPRESENTATIONS

List processing was originally introduced in their IPL programming language by Newell et al. (1957) to deal with symbol manipulation. Lists form associations of symbols which allow computer programs to build data structures of unpredictable shape and size. To handle such unpredictably shaped data structures, IPL used primitive data elements (called *cells*).

The same idea is used in LISP in the form of CONS cells. Each CONS cell is an address (a computer word) that contains a pair of pointers to other locations in computer memory.* The left portion of the cell points to the first element (the CAR) of the list. The right portion points to another CONS cell representing the remainder (the CDR) of the list. Thus, as indicated in Fig. C-1, representing a sequence of words or symbols in memory can be visualized as a binary tree structure using these memory cells. The problem of unpredictable size of data structures was solved by having a free list of memory cells that could be dynamically allocated as required.

A list is a sequence of zero or more elements enclosed in parentheses, where each element is either an atom (an indivisible element) or a list. Lists can be used to represent virtually any type of data. Lists are therefore useful for representations in such AI areas as language understanding, computer vision, and problem solving and planning.

Tree structures (used to represent search spaces) are ubiquitous in AI. A list representation for a tree structure is shown in Fig. C-2. It will be observed that the resultant representation is a list (as indicated by parentheses) consisting of elements, some of which are also lists. These nested structures are common in list representations.

Predicate logic expressions such as

$$IN(x,A) \text{ OR } IN(x,B)$$

meaning x is in A or in B, can be conveniently expressed, using prefix notation, in list form as

$$(OR \ (IN \ x \ A) \ (IN \ x \ B))$$

---

*One can thus view the basic data object in LISP to be pointers, with lists as one interpretation placed on the resultant pair structure.

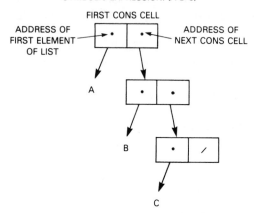

SYMBOLIC EXPRESSION: (A B C)

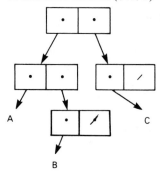

SYMBOLIC EXPRESSION: ((A B) C)

Figure C-1   Representation of list structures in memory.

C-3. LISP

C-3.1. Background

Around 1960, John McCarthy at MIT developed LISP as a practical list processing language with recursive-function capability for describing processes and problems. Since then, LISP has been the primary AI programming language—the one most used in AI research.

C-3.2. Basic Elements of LISP

All LISP programs and data are in the form of symbolic expressions (S-expressions) which are stored as list structures. LISP deals with two kinds of objects: atoms and lists. *Atoms* are symbols (constants or variables) used as identifiers to name objects which may be numeric (numbers) or nonnumeric (people, things, robots, ideas, etc.). A *list* is a sequence of zero or more elements enclosed in parentheses, where each element is either an atom or a list.

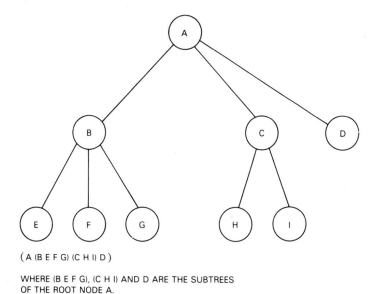

(A (B E F G) (C H I) D)

WHERE (B E F G), (C H I) AND D ARE THE SUBTREES
OF THE ROOT NODE A.

Figure C-2   List representations of a search tree.

Graham (1979, p. 226) observes: "A LISP system is a function-evaluating machine. The user types in a function and its arguments. LISP types back the result of applying the function and its arguments." For example, for addition:

User input: (PLUS 6 2)
LISP response: 8

To manipulate lists, LISP has three basic functions (related to the memory cell structure storage for lists):

CONS, to join a new first member to a list.
CAR, to retrieve the first member of a list.
CDR (pronounced coud-er), to retrieve the list consisting of all but the first member of a list.

Thus

User: (CONS 'Z '(C D E))
LISP (Z C D E)

where the single-quote symbol, ', is used to indicate that the expression following is not to be evaluated. Normally, LISP evaluates all expressions (starting with the innermost parentheses) before carrying out other operations.

User: (CAR '(John Mary X Y))
LISP: John
User: (CDR '(John Mary X Y))
LISP: (Mary X Y)

C-3.3 Variables

In LISP, the SET function assigns a value to a variable. Thus

User: (SET 'Z George)
LISP: George
User: Z
LISP: George

Atoms are used for variables in LISP. When quoted, an atom stands for itself; otherwise, LISP automatically substitutes its value during processing.

C-3.4 Defining New Functions

Programming in LISP involves defining new functions. Thus we could define SECOND (a function that retrieves the second atom of a list) by

User: (DEFUN SECOND(Y)(CAR(CDR Y)))   where Y is a dummy variable.
LISP: SECOND
User: (SECOND '(JOHN FRANK MARY JANE))
LISP: FRANK

C-3.5. Predicates

A predicate is a function that returns either NIL (false) or T (true). As a programming convenience, any none-NIL value is also considered to be true. (NIL is actually the name of the empty list.) Thus the predicate GREATERP returns T if the items in the series are in descending order:

User: (GREATERP 6 5 2)
LISP: T

C-3.6. Conditional Branching

It is often necessary in AI to use conditional branching. For example, if so and so is true, then do X; if not, if thus and so is true, then do Y; if not, do Z. The COND function in LISP has this role. Its form is

$$(COND\ (condition\ 1\ expression\ 1)$$
$$(condition\ 2\ expression\ 2)$$
$$\bullet$$
$$\bullet$$
$$\bullet$$
$$(condition\ m\ expression\ m))$$

where each condition is an expression that will evaluate to NIL or something else. The COND function evaluates the conditions in order until one of them evaluates to other than NIL. It then returns the value of the corresponding expression.

C-3.7. Recursive Functions

It is often much easier to define a function recursively—in terms of itself—than to define it as an explicit series of steps. This recursive feature is an important characteristic of LISP. A simple illustration is the factorial example (Barr and Feigenbaum, 1982, p. 6):

$$N! = \begin{cases} 1 & \text{if } N = 1 \\ N \times (N-1)! & \text{if } N > 1 \end{cases}$$

This factorial function (FACTORIAL) can be written as

$$(DEFUN\ FACTORIAL\ (N)$$
$$(COND\ ((EQUAL\ N\ 1)\ 1)$$
$$(T\ (TIMES\ N\ (FACTORIAL\ (DIFFERENCE\ N\ 1))))))$$

It should be noted that LISP programs and data are both in the same form—lists. Thus AI programs can manipulate other AI programs. This allows programs to create or modify other programs, an important feature in intelligent applications. It also allows programming aids for debugging and editing to be written in LISP, providing great interactive flexibility for LISP programmers, who can thus tailor things to suit their needs.

C-3.8. LISP Today

There are two major LISP dialects today: MACLISP, developed at MIT; and INTERLISP, developed at Bolt, Beranek, and Newman, Cambridge, Massachusetts, and Xerox Palo Alto Research Center (PARC). Both offer very similar programming environments with editing and debugging facilities. Both offer many LISP functions and optional features. The emphasis in INTERLISP has been to provide the best possible programming environments, even at the expense of speed and memory space. MACLISP has had more emphasis on efficiency, conservation of address space, and flexibility for building tools and embedding languages.

INTERLISP has been much the better supported version, with complete documentation and many users. It runs on DEC and Xerox operating systems.

Out of the need to standardize the various MACLISP dialects has evolved

Common LISP and LISP Machine LISP for personal AI computers. Common LISP appears destined to be used on most of the new personal AI machines and operating systems. Common LISP is intended to be efficient, portable, and stable.

Because of the rapid development of LISP features by the user community, other more local LISP versions (such as FRANZLISP at the University of California, Berkeley) exist at several university AI labs. A good text on LISP programming is *LISP* by Winston and Horn (1984).

C-4. PROLOG (PROgramming in LOGic)

C-4.1. History

PROLOG is a logic-oriented language developed in 1973 at the University of Marseille AI Laboratory by A. Colmerauer and P. Roussel. Additional work on PROLOG has been done at the University of Edinburgh in Great Britain. Development of PROLOG in France has continued to the present, achieving a documented system that can be run on nearly all computers (Colmerauer et al., 1981).

C-4.2. Nature of PROLOG

PROLOG is a theorem-proving system. Thus programs in PROLOG consist of "axioms" in first-order predicate logic together with a goal (a theorem to be proved). The axioms are restricted to implications written in "Horn clause" form.

A *Horn clause* consists of a set of statements joined by logical ANDs with at most one implication. Thus the form of a typical PROLOG axiom is

$$A \cap B \cap C \cap X \rightarrow Y$$

That is, A AND B AND C AND X together IMPLY Y when read declaratively. It can also be read procedurally as:

To prove Y, try to prove A AND B AND C AND X.

Looked at this way, a PROLOG program consists of a group of procedures, where the left side of a procedure is a pattern to be instantiated* to achieve the goal on the right side of the procedure.

PROCEDURE: PATTERN→GOAL

(Note the similarities of these modular rules to the if-then production rules used in constructing expert systems. It is this modularity which promotes clear, accurate, rapid programming—that is one of the reasons for PROLOG's popularity.)

Example

Find the geopolitical entities in Europe.

*Instances found that satisfy it.

Data: written as a relational data base in Horn-clause form:

$$\text{PARTOF (London, England)}$$
$$\cap \text{ PARTOF (England, Europe)}$$
$$\cap \text{ PARTOF (Boston, U.S.)}$$
$$\cap \text{ PARTOF (Tokyo, Japan)}$$

Procedures:

(1) $\text{PARTOF(X,Y)} \rightarrow \text{IN(X,Y)}$

That is, to prove that X is in Y, try to prove that X is part of Y.

(2) $\text{PARTOF(X,Y)} \cap \text{IN(Y,Z)} \rightarrow \text{IN(X,Z)}$

That is, to prove that X is in Z, try to prove that X is part of Y and that Y is in Z.

Goal (theorem to be proved):

$$\text{IN(X, Europe)}$$

That is: What X's are in Europe?

By matching the goal to the right-hand side of the first procedure, we instantiate the procedure by letting

$$\text{Europe } = \text{ Y}$$

Then matching the data to this procedure, we find

$$X = \text{England} \qquad\qquad \text{one solution.}$$

Matching the goal to the right-hand side of the second procedure, we instantiate it by letting

$$\text{Europe } = \text{ Z}$$

Now, matching the data to the two procedures, we instantiate them by letting

$$\begin{aligned} Y &= \text{England} \\ X &= \text{London} \end{aligned} \qquad \text{a second solution.}$$

Thus we have two instances:

$$\begin{aligned} X &= \text{England} \\ X &= \text{London} \end{aligned}$$

that satisfy the goal.

As indicated by the example, PROLOG solves a problem by pattern matching, which can be viewed as unification (the carrying out of instantiations) in the sense of first-order predicate logic. (PROLOG incorporates a very powerful pattern-matching mechanism.) If the pattern-matching fails as PROLOG searches through its procedures, it automatically backtracks to its previous choice point, resulting in a depth-first type of search.

The solution process starts with the system searching for the first clause whose right side matches (unifies with) the goal. Thus the search process can be guided by the programmer by choosing the order for the procedures, the data, and the goals in the clauses.

PROLOG can be considered as an extension of pure LISP coupled with a relational data base query language (as exemplified by the Horn clause form for expressing the basic data) which utilizes virtual relations (implicit relations defined by rules). Like LISP, PROLOG is interactive and uses dynamic allocation of memory.

C-4.3. PROLOG Today

PROLOG is a much smaller program than LISP and has now been implemented on a variety of computers (including microcomputers). A documented highly portable version of PROLOG has been written in France (Colmerauer et al., 1981). The execution of PROLOG is surprisingly efficient and in its compiled version it is claimed to be faster than compiled LISP. PROLOG has proved to be very popular in Europe and is now targeted as the language for Japan's Fifth Generation Computer project. PROLOG's design (and its powerful pattern matcher) is well suited to parallel search and therefore an excellent candidate for such powerful future computers incorporating parallel processing. Substantial interest in PROLOG is now arising in the United States, with some of PROLOG's features being implemented in LISP.

PROLOG's principal drawback (in its basic form) appears to be its depth-first search approach, which could be a concern in certain complex problems that tend toward combinatorial explosions in the size of the search space.

PROLOG was originally developed for natural language understanding applications, but has since found use in virtually all AI application areas.

C-5. OTHER AI LANGUAGES

A number of AI languages were developed as extensions of, improvements upon (e.g., special features for knowledge organization and search), or alternatives to, LISP. Most of these AI languages are no longer supported and have fallen into disuse. However, they were experiments that helped pave the way for the modern AI languages now in use, such as the current LISP dialects, PROLOG and POP-2 (popular in England).

Other special languages have been built for knowledge representation, knowledge base management, writing rule-based systems (such as expert systems), and for special application areas. These languages are treated in the appropriate sections of this book.

C-6. AI COMPUTATIONAL FACILITIES

C-6.1. Requirements

Good AI people are still scarce, expensive, and dedicated. It therefore behooves an organization supporting an AI group to provide the best facilities available (commensurate with cost) so as to maximize the productivity of their AI people.

Fahlman and Steele (1982) list the desirable features of an AI programming environment:

- Powerful, well-maintained, standardized AI languages
- Extensive libraries of code and domain knowledge (a facility should support the exchange of code with other AI research facilities)
- Excellent graphic displays: high resolution, color, multiple windows, quick update, and software to use all of these easily
- Good input devices
- Flexible, standardized interprocess communication
- Graceful, uniform, user-interface software
- A good editor that can deal with a program based on the individual program structure

They suggest the following:

- Sticking with the hardware and systems that major AI centers are using is important, so that the time can be spent getting work accomplished, not reinventing the wheel.
- $50,000 to $100,000 per researcher for computing facilities is appropriate.
- Your AI product be developed in the best available environment. Once developed, it can be ported to other languages and machines, as appropriate.
- Isolated machines are nearly useless. Good network interfaces, internal and external, are critical.*
- AI people spend roughly 80% of their time editing, reading, and communicating. Thus facilities for these must be excellent.

C-6.2. AI Machines

The computers used for AI research for the past several years have been primarily the DEC system-10 and DEC system-20 families of time-shared machines. These are now being superseded by the more economical DEC VAX time-shared computers and the newer personal AI machines. However, Brown (1981) saw this as a mixed blessing, as the newer machines were then still deficient in software compared to the older DEC 10s and 20s with their rich software libraries (although this situation is changing). The newer machines tend to have 32-bit words, sorely needed for address space, as most AI programs are huge.

Fahlman and Steele (1982) saw the DEC VAX, with a Berkeley UNIX operating system, as being the best time-sharing machine for AI purposes. Several LISP dialects are available. The VAX has been the choice of many universities.

*Brown (1981) sees an ARPANET link as essential for a NASA AI lab. "The ARPANET links most U.S. AI research centers and provides an electronic bulletin board, electronic mail, file transfer, and access to remote data bases, software tools and computing power."

MIT designed a personal machine specially microcoded for LISP. This MIT LISP machine has been licensed to LISP Machines Inc. (LMI) and Symbolics. Table C-1 lists some of the personal AI machines available in 1984. Several other personal machines of lesser capacity are also being offered for AI applications.

These new AI personal machines represent unusually powerful interactive exploratory programming environments in which system design and program develop together (Sheil, 1983). This is in sharp contrast to the more traditional structured programming approach, in which software program specifications are written first, with the software development following in rigid adherence to the specifications.

To further enhance the exploratory programming approach, user-friendly object-oriented programming languages have also been devised. An object (such as an airplane, or a window on a computer screen) can be encoded as a package of information with attached descriptions of procedures for manipulation of that information. Objects communicate by sending and receiving messages that activate their procedures. A class is a description of one or more similar objects. An object is an instance of a class and inherits the characteristics of its class. The programmer developing a new system creates the classes that describe the objects that make up the system and implements the system by describing messages to be sent. Use of object-oriented programming reduces the complexity of large systems. The notion of class provides a uniform framework for defining system objects and encourages a modular, hierarchical (top-down) program structure (Robinson, 1981). Smalltalk is an object-oriented language available on the Xerox machines. Flavors is available on the MIT-based LISP machines. LOOPS, being developed at Xerox PARC, is a further extension of the Smalltalk system. ROSS, at Rand Corp., is an object-oriented programming language for use in symbolic simulation of activities such as air combat.

TABLE C-1   Some Personal AI Computers Available in 1984

Machine/Company	Approximate Price	Characteristics
3600 Symbolics, Inc. Cambridge, MA	$85K	A complete redesign of the MIT LISP machine; very fast and flexible; extensive software
Lambda LISP Machines, Inc. Los Angeles, CA	$73K	A recent redesign of the MIT LISP machine; software from MIT
PERQ LN5500 PERQ Systems Corp. Pittsburgh, PA	$30K and up	A new more powerful version of the PERQ Computer. Features PERQ LISP, a superset of Common LISP.
Xerox 1100 Series Xerox Electro-Optical Systems Pasadena, CA	$30K[a] and up	A mature INTERLISP system available in several versions having different memory capacities

[a] Mainframe computer support desirable.

The new AI personal machines tend to come with interactive facilities for program development, editing, debugging, and so on. For key portions of AI programs, microcode allows key inner loops to be run very fast. This is especially important in graphics and vision programs. The efficiency of computers microcoded for AI applications and supporting large memories makes these personal computers especially attractive.

ZETALISP, derived from MACLISP, is an integrated software environment for program development and program execution on the Symbolics 3600. ZETALISP has available thousands of compiled functions, making it an exceptionally powerful and functional form of the LISP programming language. Similar capabilities are available on the LMI Lambda machines.

Interlisp-D, used on the Xerox 1100 machines, also provides a comprehensive programming environment, particularly suited for the development and delivery of expert and other knowledge-based systems.

C-7. SUMMARY AND FORECAST

It now appears that LISP dialects designed specifically for personal computers will become commonplace. It is also expected that software portability will improve substantially. PROLOG and its derivatives, now prevalent throughout Europe, will become incorporated with LISP in the United States.

It is expected that the price of very powerful AI personal computers that run LISP will rapidly drop below $50,000 as competition heats up and demand escalates. It is thus anticipated that one personal AI machine per AI person will be the standard of the future.

Parallel architectures are now being considered for future AI machines. This is especially attractive for PROLOG because its structure facilitates parallel search. Japan intends to build sequential PROLOG personal computers by 1985, featuring 10K logical inferences per second. In the 1990 time frame, Japan's Fifth Generation Computer project is projected to yield an enormously powerful AI parallel processing machine running PROLOG at 1 billion logical inferences per second (about 10,000 times more powerful than the DEC 10 on which the AI community grew up).

The rapidly increasing capability and ease of development of VLSI chips promises to move AI computing power for developed applications out of the laboratory and into the field and products as needed.

An emerging trend is the increased use of object-oriented programming to ease the creation of large exploratory programs. The use of objects is also a good way to program dynamic symbolic simulations, which will become more important as the quest for utilizing deeper knowledge accelerates and the demand for increased reliability of knowledge-based systems is pursued. Object-oriented programming also holds promise for distributed processing, as each object could be implemented on a separate processor in a linked network of processors.

Finally, it is anticipated that the AI exploratory software development approach will slowly infuse conventional software practices.

REFERENCES

Barr, A., and Feigenbaum, E. A. (Eds.), *The Handbook of Artificial Intelligence*, Vol. I. Los Altos, CA: W. Kaufmann, 1981.

Barr, A., and Feigenbaum, E. A. (Eds.), *The Handbook of Artificial Intelligence*, Vol. II. Los Altos, CA: W. Kaufmann, 1982.

Brown, D. R., "Recommendations for an AI Research Facility at NASA/GSFC," SRI Project 2203, SRI International, Menlo Park, CA, Oct. 1981.

Colmerauer, A., Kanoui, H., and Van Canegham, M., "Last Steps toward an Ultimate PROLOG," *IJCAI-81*, Vancouver, Canada, Aug. 1981, pp. 947–948.

Fahlman, S. E., and Steele, G. L., "Tutorial on AI Programming Technology: Language and Machines," *AAAI-82*, Pittsburgh, PA, Aug. 16, 1982.

Graham, N., *Artificial Intelligence*. Blue Ridge Summit, PA, TAB Books, 1979.

Newell, A., Shaw, J. C., and Simon, H. A., "Programming the Logic Theory Machine," *Proceedings of the Western Joint Computer Conference*, 1957, pp. 230–240.

Robinson, D., "Object-Oriented Software Systems," *Byte*, Aug. 1981, pp. 74–86.

Sheil, B., "Power Tools for Programmers," *Datamation*, Feb. 1983, pp. 131–144.

Winston, P. H., and Horn, B. K. P., *LISP.*, 2nd ed. Reading, MA: Addison-Wesley, 1984.

D

A BRIEF HISTORY OF AI COVERING
ITS RISE, FALL, AND REBIRTH

D-1. THE FIRST 15 YEARS

In 1956, ten scientists convened a conference at Dartmouth College from which emerged the present field of AI. The gist of the predictions made by those scientists was that in 25 years, we would all be involved in recreational activities, while computers would be doing all the work. In 1981, at the International Joint Conference on AI in Vancouver, Canada, a panel of five of these same scientists recalled that conference and their overoptimistic forecasts.

In 1956 it was assumed that intelligent behavior was based primarily on smart reasoning techniques and that bright people could readily devise ad hoc techniques to produce intelligent computer programs.

Figure D-1 lists some of the key AI activities during the first 15 years.* The major initial activity involved attempts at machine translation. It was thought that natural language translation could be readily accomplished using a bilingual dictionary and some knowledge of grammar. However, this approach failed miserably because of factors such as multiple word senses, idioms, and syntactic ambiguities. A popular story is that the saying "The spirit is willing but the flesh is weak," when translated into Russian and back again into English, came out "The wine is good but the meat is spoiled." Schwartz (1980, p. 27) reports that "twenty million dollars of mechanical translation brought results so disappointing, that . . . by 1967 opinion had soured so dramatically that the National Academy of Sciences all but created a tombstone over the research." In fact, it has only been recently that substantial work in mechanical language translation has reappeared.

*An interesting history of the early years of AI is given by McCorduck (1979).

Activities

- Attempts at Machine Translation

- ELIZA—Key Word and Template Matching

- Symbolic Integration

- Game Playing—Checkers, Chess

- Pattern Recognition

- Computational Logic

- General Problem Solver

Lessons Learned

- AI Much More Difficult Than Expected

- Heuristic Search Required To Limit Combinatorial Explosion

- Lack of Contextual Knowledge Severely Limits Capability

- Expectation Is a Human Characteristic of Intelligence

- Difficult To Handle a Broad Domain (e.g., Common Sense)

Figure D-1   Condensed history of AI: 1956–1970.

Weizenbaum (1966) at MIT designed a natural language understanding program that simulated a nondirective psychotherapist. The program (ELIZA) bluffed its way through the interaction by picking up on key words and providing stock answers. When it did not find a recognizable key word, it would select a reply such as "Please continue." Although Weizenbaum wrote the program in part to show how ridiculous it was to expect true natural language understanding by a machine, the program nevertheless became popular and some of its basic techniques are used in commercial natural language interfaces today.

In 1961, Slagle at MIT devised a heuristic computer program to do symbolic integration. This proved to be the forerunner of a successful series of symbolic mathematical programs culminating in MACSYMA, in use at MIT today and available over ARPANET to other AI researchers, and also available commercially.

Game playing was also one of the early areas of AI research, with Samuel's (1963) work at IBM on machine learning in checkers proving to be one of the early successes. Solving puzzles was another area of early success in AI, leading to the development of problem-solving techniques based on (1) search and (2) reducing difficult problems into easier subproblems.

Early work in vision involved image processing and pattern recognition (which was concerned with classifying two-dimensional patterns). Pattern recognition split off from AI and became a field in itself, but now the two disciplines have become much more unified.

The pioneering work in computer vision was Roberts' (1965) program designed to understand polyhedral block scenes. This program found the edges of the blocks using the spatial derivatives of image intensity, and from the resulting edge elements produced a line drawing. It then utilized simple features, such as the numbers of vertices, to relate the objects in the line drawing to stored three-dimensional models of blocks. The resulting candidate model was then scaled, rotated, and projected onto the line drawing to see if the resultant match was adequate for recognition.

Another important area was computational logic. Resolution, an automatic method for determining if the hypothesized conclusion indeed followed from a given set of premises, was one of the early golden hopes of AI for universal problem solving by computer. Using resolution, Green (1969) devised a general-purpose, question-answering system, QA3, that solved simple problems in a number of domains, such as robot movements, puzzles, and chemistry. Unfortunately, resolution, although it guarantees a solution, devises so many intermediate steps that turn out not to be needed for the final solution that for large problems its use results in a combinatorial explosion of search possibilities.

Another approach, originally thought to have broad applicability, was the General Problem Solver (GPS) devised by Newell et al. (1960). The generality resulted from GPS being the first problem solver to separate its problem-solving methods from knowledge of the specific task currently being considered. The GPS approach was referred to as "means-ends analysis." The idea was that the differences between the current problem state and the goal state could be measured and classified into types. Then, appropriate operators could be chosen to reduce these differences, resulting in new problem states closer to the goal states. This procedure would then be iteratively repeated until the goal was reached. The series of operators used would then form the solution plan. Unfortunately, classifying differences and finding appropriate operators turned out to be more difficult than expected for nontrivial problems. In addition, computer running times and memory requirements rapidly become excessive for the more difficult problems.

AI proved much more difficult than originally expected. By 1970, AI had had only limited success. Natural language translation had already collapsed. "Toy" problems or well-constructed problems such as games proved tractable, but real complex problems proved to be beyond the techniques thus far devised, or resulted in combinatorially explosive search that exceeded the computer capabilities then current. Similarly, real-world computer vision efforts tended to be overwhelmed by the noise and complexities in real scenes.

In 1971, J. Lighthill of Cambridge University was called upon by the British government to review the AI field. The Lighthill report (1972) found that "in no part of the field have the discoveries made so far produced the major impact that was promised." Further, he found that respected AI scientists were then predicting that "possibilities in the 1980's include an all-purpose intelligence on a human-scale knowledge base; that awe-inspiring possibilities suggest themselves based on machine intelligence exceeding human intelligence by the year 2000"—the same sort of forecasts as were made 15 years earlier. Lighthill saw no need for a separate AI field and found no organized body of techniques that represented such a field. He felt that the work in automation and computer science would naturally come together to bridge whatever gap existed. The Lighthill report eventually brought work in AI in England to a virtual halt and cast a pall over AI work in the United States.

However, the AI efforts of the 1950s and 1960s were not without merit. A great deal was learned about what really had to be done to make AI successful. It was found that expectation is a human characteristic of intelligence. That perception, both visual and in language, is based on knowledge, models, and expectations of the perceiver. Thus communication via language was found to be based on shared knowledge between the participants and that only cues are needed to actualize the models (in the receiver's head) from which to construct the complete message.

Thus in attempting communication or problem solving, lack of contextual knowledge was found to limit capability severely. Reasoning techniques alone proved inadequate. Knowledge is central to intelligence. Lacking this knowledge, it is difficult to handle a broad domain. An example is "common sense," found to be elementary reasoning based on massive amounts of experiential knowledge.

It was also found that heuristics are necessary to guide search to overcome the combinatorial explosion of possible solutions that pervade complex problems—for each time a decision is made, new possibilities are opened up.

D-2. THE DECADE OF THE 1970s

As indicated in Fig. D-2, in the 1970s AI researchers began to capitalize on the lessons learned. New knowledge representation techniques appeared. Search techniques began to mature. Interactions with other fields, such as medicine, electronics, and chemistry, took place. Feasible approaches were demonstrated for language processing, speech understanding, computer vision, and computer programs that could perform like experts.

SHRDLU was a natural language program at MIT devised by Terry Winograd (1972) to interface with an artificial "blocks world." It was the first program to deal successfully in an integrated way with natural language by combining syntactic and semantic analysis with a body of world knowledge.

From 1971 through 1976, ARPA sponsored a five-year speech understanding program. Hearsay II at Carnegie-Mellon University was a winner, being able to understand sentences, with 90% accuracy, from continuous speech based on a 1000-word vocabulary. (The "blackboard" system architecture, devised for Hearsay II to deal with multiple knowledge sources, has since found use in other AI applications.) A compiled network architecture system called HARPY, which handled the same vocabulary as Hearsay II, was able to achieve 95% accuracy. (A more detailed review is given in Chapter 10.)

Activities

- Feasible Approaches Demo'd for:
  - Language Processing
  - Computer Vision
  - Expert Systems
  - Speech Understanding

- New Knowledge Representation Techniques Appear

- Search Techniques Begin To Mature

- Interaction with Other Fields Takes Place

Lessons Learned

- Knowledge Central to Intelligence

- Future Complex Systems Proved Feasible

Figure D-2   Decade of the 1970s.

At SRI, Gleason and Agin (1979) developed the SRI Vision Module as a prototype system for use in industrial vision systems. This system, which used special lighting to produce a binary image (silhouette) of an industrial workpiece, was able to extract edges by a simple continuous scan process, and was to prove the basis for several sophisticated commercial vision systems.

In the 1970s, following an earlier successful effort called DENDRAL, a variety of prototype computer programs—called expert systems—designed to capture and utilize the expertise of a human expert in a narrow domain (such as medical diagnosis, crystallography, electrical circuitry, prospecting, etc.) made their appearance. MYCIN, a medical diagnosis and treatment consultant devised by Shortliffe (1976) at Stanford University, has been one of the most publicized.

Thus the 1970s found the AI research community developing the basic tools and techniques needed, and demonstrating their applicability in prototype systems. Future complex systems were proved feasible. The emphasis on knowledge, as essential to intelligence, led to the subfield of "knowledge engineering" associated with the building of expert systems.

D-3. 1980 TO THE PRESENT

The decade of the 1970s set the framework from which the successes of the 1980s emerged. In the 1980s, expert systems proliferated. Hundreds of prototype expert systems were devised in such areas as medical diagnosis, chemical and biological synthesis, mineral and oil exploration, circuit analysis, tactical targeting, and equipment fault diagnosis.

But the big news of the 1980s (see Fig. D-3) is that AI has gone commercial. AI companies (founded mostly by AI researchers) have formed to exploit applications. Computer, electronic, oil, and large diversified companies have set up AI groups. The military has also joined the fray, setting up their own AI groups and seeking early applications. The U.S. Defense Science Board views AI as one of the technologies that has the potential for an order-of-magnitude improvement in mission effectiveness.

In the expert systems area, DEC reports that R1—a system designed to configure VAX computer systems—is already saving them some $20 million a year. MOLGEN—a system for planning molecular genetic experiments—is in regular commercial use. Shlumberger—a multibillion-dollar oil industry advisory company—seeing AI as a key to the company's growth in the 1980s, has established four separate

Activities

- Expert Systems Proliferate

- AI Goes Commercial

  - Expert Systems: RI, DIP-METER ADVISOR, MOLGEN

  - Natural Language Front Ends—INTELLECT

  - Speech Output—Speak and Spell

  - Vision Systems

  - AI Groups and Companies Form To Exploit Applications

  - LISP Machines Become Available

- AI Technology Becoming Codified

  - AI Handbook

  - Individual Technology Texts: Natural Language, Vision, etc.

  - NBS/NASA Overviews

Conclusions

- AI Tools and Systems Become Available

- Logic Systems (Heuristically Guided) Reemerge—PROLOG

- AI Techniques Sufficiently Perfected for Early Applications

Figure D-3 1980–present.

AI groups. The Connecticut group has already created the expert system Dip-Meter Advisor, to evaluate oil-drilling core samples.

In natural language front ends, over a dozen systems are now commercially available. Intellect, from Artificial Intelligence Corporation, already had several hundred installations by 1984.

Highlighted by Texas Instruments' Speak and Spell, many commercial speech output systems have appeared. Limited speech recognition systems are also on the market, some using signal processing rather than AI techniques.

Hundreds of companies are now involved in computer vision systems, with dozens of commercial products already on the market for simplified vision applications.

Personal computers that are specially designed to run LISP—the list processing language favored by the U.S. AI community—are now commercially available from several companies.

The other indication that AI has now emerged as a viable discipline is that the existing AI technology is now becoming codified and therefore made broadly available to everyone, not just the core group of several hundred researchers of the 1970s.

ARPA sponsored a three-volume *Handbook of Artificial Intelligence* which was published in 1981 and 1982. Individual technology texts—in vision, natural language, expert systems, and LISP—are beginning to appear in numbers.

NASA sponsored a NBS/NASA set of overviews in artificial intelligence and robotics, entitled *An Overview of Artificial Intelligence and Robotics* and containing the following volumes:

Vol. I: *Artificial Intelligence*, NASA TMs 85836, 85838, 85839, 1983

Vol. II: *Robotics*, NBSIR 82-2479, Mar. 1982

Vol. III: *Expert Systems*, NBSIR 82-2505, May 1982 (rev. Oct. 1982)

Vol. IV: *Computer Vision*, NBSIR 82-2582, Sept. 1982

Vol. V: *Computer-based Natural Language Processing*, NBSIR 83-2687, Apr. 1983. NASA TM 85635, Apr. 1983

Computer software tools for structuring knowledge and constructing expert systems are proliferating.

In 1982, the Japanese officially began a 10-year, $500 million research project to create a *fifth-generation computer*. The main features of this computer are that it is to have (1) intelligent interfaces (speech, text, graphics, etc.), (2) knowledge base management, and (3) automatic problem-solving and inference capabilities. All these capabilities are predicated on the use of AI techniques. The machine itself is visualized as a non-von Neumann computer featuring parallel processing and having the capability of 1 billion logical inferences per second.

The Japanese are considering the European AI language PROLOG (Programming in Logic) as the basis for their machine. Using PROLOG, logic problem-solving systems (heuristically guided) are reemerging (from the earlier failure of pure resolution) to handle complex problems.

With the advent of the Japanese Fifth-Generation Computer project, European nations, such as France and Great Britain, as well as the United States, are putting renewed effort into their AI activities (Warren, 1982).

In summary, then, we can conclude that AI tools and systems are now becoming available and that AI techniques are now sufficiently perfected for early applications. Further, the importance of AI is being recognized internationally and substantial sums of money in the United States and abroad are now being committed to developing AI applications.

REFERENCES

Gleason, G. J., and Agin, G. J., "A Modular System for Sensor-Controlled Manipulation and Inspection," *Proceedings of the Ninth International Symposium of Industrial Robots*, SME and RIA, Washington, DC, 1979, pp. 57–70.

Green, C. C., "The Application of Theorem-Proving to Question Answering Systems," *IJCAI-1*, 1969, pp. 219–237.

Lighthill, J., "Artificial Intelligence: A General Survey," Scientific Research Council of Britain, SRC 72-72, Mar. 1972.

McCorduck, P., *Machines Who Think*. San Francisco: W. H. Freeman, 1979.

Newell, A., Shaw, J. C., and Simon, H. A., "A Variety of Intelligent Learning in a General Problem Solver," in *Self Organizing Systems*, M. C. Yovits and S. Cameron (Eds.). Elmsford, NY: Pergamon Press, 1960, pp. 153–189.

Roberts, L., "Machine Perception of Three-Dimensional Solids," in *Optical and Electro-Optical Information Processing*, J. Tippitt (Ed.). Cambridge, MA: MIT Press, 1965, pp. 159–197.

Samuel, A. L., "Some Studies in Machine Learning Using the Game of Checkers," in *Computers and Thought*, E. A. Feigenbaum and J. Feldman (Eds.). New York: McGraw Hill, 1963.

Schwartz, R. D., "Refocus and Resurgence in Artificial Intelligence," *IEEE Proceedings of the National Aerospace and Electronic Conference, NACON-1980*, Dayton, OH, May 20–22.

Shortliffe, E. H., *Computer-Based Medical Consultations: MYCIN*. New York: American Elsevier, 1976.

Slagle, J. R., "A Heuristic Program That Solves Symbolic Integration in Freshman Calculus: Symbolic Automatic Integrator (SAINT)," Rep. 5G-001, Lincoln Lab., MIT, Cambridge, MA, 1961.

Warren, D. H. D., "A View of the Fifth Generation and Its Impact," *AI Magazine*, Vol. 3, No. 4, Fall 1982, pp. 34–39.

Weizenbaum, J., "Eliza—A Computer Program for the Study of Natural Language Communication between Man and Machine," *Communications of the ACM*, Vol. 9, 1966, pp. 36–45.

Winograd, T., *Understanding Natural Language*. New York: Academic Press, 1972.

E

THE PRINCIPAL PARTICIPANTS

Originally, AI was principally a research activity, the principal centers being Stanford University, MIT, Carnegie-Mellon University, SRI, and the University of Edinburgh in Scotland. Research successes during the 1970s encouraged other universities to become involved.

In the 1980s it became apparent that AI had a large commercial and military potential. Thus existing large computer, electronic, and multinational corporations, as well as some aerospace firms, started forming AI groups. Shlumberger, a multibillion-dollar oil exploration advisory firm, was the first to pursue AI in a big way. They now have AI groups in France, Connecticut, Texas, and California.

In 1980, the Navy committed itself to building an AI laboratory at Bolling AFB, Washington, DC, to help transfer AI to Navy applications from research at the universities. The lab has several dozen people, including Navy personnel and visiting scientists, and is now undergoing transition to a tri-service lab. By 1982, the Army and the Air Force also decided to form AI organizations and are now in the process of doing so.

In response to a perceived market in natural language processing, computer vision, and expert systems, small new AI companies began to form, headed by former (and present) university researchers. Several dozen such companies now exist. The computer science departments at major universities have also recently become involved, so that AI courses and beginning AI research is now evident at many universities.

Abroad, France and Great Britain have now joined Japan in evidencing major concern. The first major commitment to AI has been by Japan, which has initiated a 10-year, $500 million program to develop a fifth-generation computer. This computer is to incorporate a parallel processing architecture, natural language inter-facing, knowledge base management, automatic problem solving, and image under-standing as the basis for a truly fast, intelligent computer. In the United States, a new cooperative organization—Microelectronics Computer Technology Corporation (MCC)—made up of U.S. computer and electronics manufacturers, has recently been formed to be a sort of American version of the Japanese Fifth-Generation Computer research project. A major program (The Strategic Computing System) focusing on military applications is being funded by DARPA.

Thus the AI research sponsored by DARPA, NIH, NSF, ONR, and AFOSR for the past two decades has now spawned such a burgeoning AI community that it is no longer an easy task to list all those involved. (A list of participants in each of the application areas is given in the associated volumes in this series.) However, Table E-1 provides an indication of the current principal players. These are given by application area, as most research efforts initially have a specific application area as a focus, with the results of the research usually being generalized later to cover a broader area.

TABLE E-1 Principal Participants in AI

Type of Group	Expert Systems	Computer Vision	Natural Language Processing
1. Universities	Stanford Univ. MIT Carnegie-Mellon Univ. Rutgers Univ.	Carnegie-Mellon Univ. Univ. of Maryland MIT Stanford Univ. Univ. of Rochester Univ. of Massachusetts	Yale Univ. Univ. of California, Berkeley Univ. of Illinois Brown Univ. Stanford Univ. Univ. of Rochester
2. Nonprofit organizations	SRI Rand JPL MITRE	JPL SRI ERIM	SRI
3. U.S. government	Navy AI Lab., Washington, DC NOSC, San Diego	NBS, Washington, DC	Navy AI Lab.
4. Diversified industrial corporations	Fairchild Camera and Instrument Schlumberger-Doll Research Hewlett-Packard Bell Labs Hughes IBM Digital Equipment Corp. GM Martin Marietta Texas Instruments TRW Xerox PARC AMOCO United Technologies Corp. Atari	GE Hughes GM Westinghouse	Bolt, Beranek and Newman IBM TRW Burroughs SDC Hewlett-Packard Martin Marietta Texas Instruments Bell Labs. Sperry Univac Lockheed Electronics Corp.

TABLE E-1 Principal Participants in AI

Type of Group	Expert Systems	Computer Vision	Natural Language Processing

5. New AI Companies	Grumman Aerospace Corp.
	Lockheed Palo Alto
	Westinghouse Electric Corp.
	Boeing Computer Services
	FMC
	
	AIDS, Mt. View, CA
	Applied Expert Systems,
	  Cambridge, MA
	Brattel Research Corp.,
	  Boston, MA
	Daisy, Sunnyvale, CA
	Intelligent Software,
	  Van Nuys, CA
	Jacor, Alexandria, VA
	Kestrel Institute,
	  Palo Alto, CA
	Smart Systems Technology,
	  Alexandria, VA
	Systems Control, Inc.,
	  Palo Alto, CA
	Teknowledge, Inc.,
	  Palo Alto, CA
	IntelliCorp, Palo Alto, CA
	Software Architecture and
	  Engineering, Arlington, VA
	Syntelligence,
	  Sunnyvale, CA
	Tektronix,
	  Beaverton, OR
	The Carnegie Group, Inc.,
	  Pittsburgh, PA
		Automatix, Inc.,
		  Billerica, MA
		Machine Intelligence Corp.,
		  Sunnyvale, CA
		Octek, Burlington, MA
			AIC, Waltham, MA
			Cognitive Systems Inc.,
			  New Haven, CT
			Symantec, Cupertino, CA
			Computer Thought,
			  Plano, TX
			Weidner Communications
			  Corp., Provo, UT

6. Computer Manufacturers
   and Researchers

   Inference Corp.,
     Los Angeles, CA
   LISP Machines, Inc., Cambridge,
     MA, Culver City, CA
   Symbolics, Cambridge, MA
   PERQ Systems Corp.
     Pittsburgh, PA
   Digital Equipment Corp.,
     Hudson, MA
   Xerox, Pasadena, CA
   Daisy, Sunnyvale, CA
   Tektronix, Inc., Beaverton, OR
   Bolt, Beranek and Newman,
     Cambridge, MA
   MCC, Austin, TX (U.S. fifth-
     generation computer research
     consortium)

7. Major Foreign Participants

   Japan (Fifth-generation computer)
   Electromechanical-Technology Lab,
     Tsukiba
   Fujitsu-Fanuc Ltd., Kawasaki
   Hitachi Ltd., Tokyo
   Mitsubishi Elec. Corp., Tokyo
   Nippon Electric Co. Ltd., Tokyo
   Nippon Tele and Tele Corp., Tokyo
   Great Britain
   Imperial College, London
   Univ. of Edinburgh, Scotland
   Univ. of Sussex
   Intelligent Terminals Ltd.
   France:
   Univ. of Marseilles, Marseilles
   Italy:
   Univ. of Milan

   } Fifth-generation computer

F

SPEECH SYNTHESIS

F-1. INTRODUCTION

Speech synthesis—speech output from a computer—is an emerging technology whose products are already becoming commonplace. Although the present market for these devices is still small, the future looks very bright.

Speech synthesis is not normally considered an AI topic, although it is sure to play an important part in many future AI systems, particularly when coupled with speech understanding. One may very well consider these synthesis systems, which employ rules (often heuristic) for deriving speech from stored speech elements, as an example of an "expert system on a chip."

F-2. WHY SYNTHESIS

One approach to making available speech when needed is to record the speech and play it back as required. The disadvantage is that mechanical devices are often unreliable and the ability to generate new sentences from stored words is quite limited because of access time, and therefore unsuitable for most computer-based applications.

A more reliable approach is to use digital sound recording techniques, enabling speech to be stored in solid-state memories having no moving parts to break down. The disadvantage is that an enormous amount of storage is required—on the order of 50,000 bits per second of digital speech (at the typical speaking rate of 150 words per minute). However, if words are represented by the digital code for their letters, the same information requires only about 100 bits per second of speech. This two- to three-order-of-magnitude difference highlights the importance of speech compression for any digital representation of speech, not only to save storage requirements, but also to vastly reduce the bandwidth required for electronic speech transmission. All speech synthesis methods use some form of speech compression.

Speech synthesis serves three basic purposes:

1. Recreating speech from a compressed speech representation
2. Generating speech from stored speech elements such as by concatenating representations for words
3. Generating speech from text

The first purpose is associated with minimizing storage or transmission bandwidth requirements, the second with creating speech from stored components under microprocessor or computer control, and the third with reading machines and computer-human interaction.

An indication of applications of speech synthesis is given in Table F-1.

TABLE F-1 Applications of Speech Synthesis

Military
    Operation of military equipment
    Warnings
    Reminders
    Service and operation aids
    Trainers and simulators
    Secure communications

Computer
    Communication by computers to users

Consumer
    Talking appliances
    Teaching devices
    Toys
    Talking typewriters and calculators
    Talking watches
    Automobile warning devices, reminders, and annunciators for instruments
    Devices for the blind
    Communication for the speech handicapped

Telecommunications
    Synthesized telephone messages
    Speech compression for "store and forward," to reduce communication costs
    Vocal delivery of electronic mail

Industrial
    Speaking instruments
    Speaking cash registers
    Alarm systems
    Automated office equipment
    Industrial process control
    Station and floor announcers for trains, buses, elevators, etc.
    Systems operations where the operators have their visual attention elsewhere
    Emergency warning devices for airplanes, machines, etc.
    Control room annunciators for sensors
    Text readers
    Data entry (with vocal verification)

F-3. HUMAN SPEECH

As many speech synthesizers actually employ an approximate simulation of the human speech production mechanism, it is helpful to briefly review human speech and its generation. Human speech consists basically of a combination of vocal sounds, such as vowels; fricative sounds, such as f, th, or sh; and plosive or stop consonant sounds, such as b and d.

The human vocal tract can be considered as an acoustic tube terminated at one end by the vocal cords and at the other end by the lips. This resonant tube has a side branch—the nasal resonator—separated by a flap called the velum.

Voiced sounds are produced by forcing air from the lungs past the tensed vocal cords, which are thus forced to vibrate, emitting puffs of air into the vocal tract. (The puff frequency—about 100 hertz in males, 200 hertz in females—is a function of the vocal cord size and tenseness.) These puffs of air excite the vocal tract, stimulating their resonant (formant) frequencies. Most of the resulting sound energy is contained in these resonant responses, the frequency of which can be varied by changing the shape of the vocal tract by moving the lips, jaw, or tongue.

Fricative sounds occur when a constriction in the vocal tract leads to turbulent airflows after the constriction.

Plosives are generated by briefly closing the vocal tract until pressure builds up and then releasing the pressure.

F-4. ELECTRONIC SIMULATION OF THE
SPEECH MECHANISM

The three basic human speech sounds can be electronically simulated as follows, as illustrated by the Computalker Consultants Model CT-1* synthesizer shown schematically in Fig. F-1. Voiced sounds can be simulated by passing energy from a variable

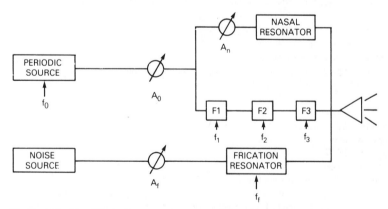

Figure F-1 Simplified diagram of Computalker CT-1 parametric synthesizer. (From Sherwood, 1979 © 1979 IEEE. Reprinted with permission.)

*No longer in production, but the Philips speech chip essentially does the same thing.

periodic source, corresponding to the vocal cord puffs, through a series of variable filters (f_1, f_2, f_3), corresponding to the vocal tract resonances (formants). Plosive sounds are produced the same way but require rapid changes in the amplitude parameters A_o and A_n. Fricative sounds are produced by passing white noise through a variable filter (f_f). Some sounds, such as v and z, are produced using both the periodic and noise mechanisms.

Using this approach, human speech can be simulated by controlling the frequency parameters (f_i) and the amplitude parameters (A_i) over time. Some variant of this basic method, referred to as parametric coding, is used in all speech synthesizers that simulate human speech production.

F-5. SYNTHESIS IN SPEECH COMPRESSION AND REGENERATION

Synthesis has the role of regeneration in speech compression schemes (associated with speech storage or minimal-bandwidth speech transmission).

There are two basic speech compression techniques: frequency-domain analysis (parametric coding as discussed in Section F-4) and time-domain analysis. Frequency-domain methods tend to dominate commercial speech synthesis, but time-domain analysis has become important for limited-vocabulary word synthesis.

The frequency-domain approach analyzes the incoming speech to be compressed and generates the parameters needed for regenerating the signal using an electronic simulation of the vocal tract. In some cases, these parameters may be further compressed for reduced storage. Speech is generated by inverting the process, as indicated in Fig. F-2.

Time-domain analysis is characterized by waveform compression techniques. Waveform digitization coding, researched extensively by Bell Labs., takes the original waveform of spoken words and compresses them using a complicated algorithm. The final compressed waveform is stored as bits in memory for later reconstruction

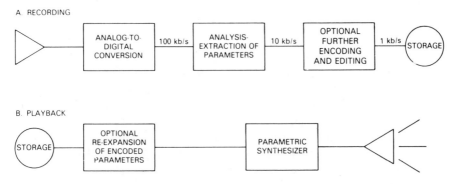

Figure F-2 Recording and reproduction of speech using a compressed-speech system. (From Sherwood, 1979 © 1979 IEEE. Reprinted with permission.)

of the original waveform. Although generally producing better-sounding speech than parametric coding, waveform digitization coding requires two to four times as much storage as that needed for parametric coding.

F-6. PARAMETRIC CODING SCHEMES

F-6.1. Introduction

All frequency-domain compression techniques employ some sort of electronic model of the human vocal tract. Thus all have one or more filters to simulate vocal tract resonances and periodic and noise energy sources, and are controlled by varying the parameters associated with pitch, loudness, and filter frequencies.

F-6.2. Formant Coding

This is a straightforward approach to controlling an electronic model of the vocal tract by controlling the tunable filters using parametric signals that represent the formant (vocal tube resonant) frequencies, such as those shown in Fig. F-1. As the formant frequencies change relatively slowly, the parameters need to be updated relatively infrequently, thus allowing data compression.

F-6.3. Linear Predictive Coding (LPC)

LPC, pioneered by Texas Instruments for Speak and Spell, is a form of formant coding which allows further compression of the parameters. As the formant frequencies tend to change slowly, current samples are predicted from weighted linear combinations of previous samples. LPC's clever prediction approach and the use of an ingenious lattice filter greatly simplifies the synthesis circuitry. The resulting system can be stored on a single chip and produces high-quality natural-sounding speech.

F-6.4. PARCOR

PARCOR (partial correlation), utilized by Japanese manufacturers, is a variant of LPC. LPC extrapolates from a series of formant samples to predict following formant frequencies. Although most speech patterns change slowly, plosive and fricative sounds involve rapid changes. PARCOR makes LPC more sensitive to sudden changes by giving greater emphasis to the correlation between adjacent parametric samples and less to the longer-term patterns. However, there appears to be little resultant subjective differences in observed speech quality between the two approaches.

F-6.5. Line Spectrum Pair (LSP)

NTT (Nippon Telephone and Telephone Public Corp.), which developed PARCOR, has come up with LSP, an approach allowing still further compression. LSP defines the boundary conditions for the individual formant frequencies as those corresponding to the open and closed vocal tract. NTT claims that for a complete system, some 40% more compression can be achieved with LSP than with PARCOR, while maintaining nearly the same speech quality.

F-6.6 Parametric Waveform Coding (PWC)

PWC is another variant of LPC, as used by Centigram's Voice Ware system to produce vocabularies for the Lisa Speech Board.* PWC uses a variable-length slice of waveform to produce the linear prediction coefficients. Each slice (about 20 milliseconds in length) corresponds to a "glottal event," the event associated with each puff of air passing through the vocal tract. Voice Ware uses an array processor to determine 13 linear prediction coefficients for each glottal event. To synthesize speech, the Lisa Speech Board uses these coefficients and the lengths of the events to recreate speech waveforms as in other LPC synthesizers. The PWC approach tends to yield more natural speech than the simpler LPC systems, but requires a higher data rate.

F-7. WAVEFORM CODING SCHEMES

F-7.1. ADPCM

Digitized speech at an 8-kHz sampling rate results in 32,000 bits per second (bps) for a 4-bit sampling size using the adaptive differential pulse code modulation (ADPCM) proposed as the worldwide preferred method of digitized voice telephone signals for long-distance transmission. In ADPCM, the digitized speech is encoded in terms of the amplitude differences between adjacent samples. These differences are adaptively encoded in terms of quantization level (a function of the previous quantization level and the previous PCM value). A close relative of ADPCM is CVSD (continuous variable-slope delta modulation).

F-7.2. Mozer's Waveform Coding

Although ADPCM is suitable for telephone transmission, its high bit rate is unsuitable for stored speech synthesis. A scheme by Forrest Mozer of the University of California is a variation of ADPCM which provides substantial further

---

*No longer in production.

compression. This technique has been incorporated into the National Semiconductors Corporation's Digitalker. Mozer's approach is to:

1. Analyze the waveform to detect short periods with little change. The waveforms for these periods are then replaced with identical waveforms.
2. Fourier-analyze the signal and adjust the phase angle of each Fourier component to produce a symmetrical waveform and then discard half.
3. Discard low-amplitude portions of the waveform which are not heard by the ear.
4. Employ ADPCM to reduce data further.

The net result of these actions is more than a 40:1 reduction in the data that need to be stored compared with the data in direct digitization. To produce speech the process is inverted. Although these resultant signals look little like the original, the result is very good speech reproduction.

F-8. CODING THE WORDS TO BE STORED

Although the schemes discussed thus far provide a huge amount of reduction in the storage required, generating the required custom vocabulary in terms of the stored parameters requires hand tailoring by an expert. As yet, there is no acceptable automatic mechanism for directly converting speech into satisfactory storage elements for encoding schemes that provide high data compression. (ADPCM is automatic. Parametric schemes can be automated with small residual errors.)

Developing the vocabulary for the Mozer waveform coding used in National Semiconductor's Digitalker takes about 1 hour of processing per word. It involves working with the data compression and zero phase-encoding algorithms that produce the stored bit patterns, making it very difficult for users to program their own custom vocabularies (Ciarcia, 1983).

To enable users to develop their own custom vocabularies for their products, when large vocabularies are required, Centigram Corp. has offered as a product their Voice Ware development system. With it, users can input a tape-recorded voice to a digitizer that supplies a 4800-bps data stream to a microprocessor-based CRT-terminal workstation. The station converts the signal into parametric waveform coding (PWC). The user can then edit the messages, combine them into files, and feed them back through the Lisa synthesizer to hear how they sound. If the sound is unsatisfactory, particularly for concatenated phrases, the phrases can be re-recorded to achieve the desired continuity and balance.

In general, for synthesizer users requiring a small custom vocabulary, it is customary for them to contract with the synthesizer manufacturer or other development source for the words required. This cost is in the order of $100 per word for LPC chips.

F-9. GENERATING SPEECH FROM TEXT

English has some 40 basic speech sounds called phonemes, corresponding to 16 vowel sounds, 6 stops, 8 fricatives, 3 nasals (such as "ng"), 4 liquids/glides (such as the "l" in "lice") and 3 others (such as the "ch" in "church"). These sounds vary somewhat depending upon how they are combined into words or used in speech. These phoneme variations are called allophones. (Texas Instruments developed a set of 128 allophones to characterize English speech.) Allophones and the rules to string them together can be stored in computer memory chips. The first text-to-speech system used a phonemic synthesizer (Votrax). Votrax utilized a hard-wired phonemic-to-parameter converter which then fed a formant synthesizer to create speech. A simplified text-to-speech system schematic is given in Fig. F-3.

A highly intelligible state-of-the-art speech synthesizer, the Speech Plus Prose 2000, utilizes a generation approach consisting of five serial processes: (1) text normalization, (2) phonemics, (3) allophonics, (4) prosodics, and (5) parameter generation. For words not in the exceptions lexicon, the phonemics process is implemented as a real-time expert system consisting of a small rule interpreter and an ordered set of about 400 context-sensitive rules.

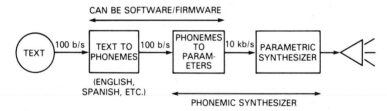

Figure F-3 Text-to-speech synthesis. (From Sherwood, 1979 © 1979 IEEE. Reprinted with permission.)

F-10. STATE OF THE ART

Elphick (1981, p. 42) notes that

> Most commercial synthesizers, especially low-cost ones used for consumer products, derive their speech elements from recordings of actual human speech. The recorded speech patterns are compressed, and the speech is disassembled into a vocabulary of small elements for later reassembly into messages.

High-quality speech by phoneme synthesizers has been achieved in research systems but not in commercial systems. The most natural commercial speech synthesizers use the waveform approach.

Figure F-4 is an indication of speech quality versus bit storage requirements for the various synthesis techniques. Thus far, in industrial applications, only short messages are practical, as prolonged listening to synthetic speech tends to fatigue the operators (Andreiev, 1981).

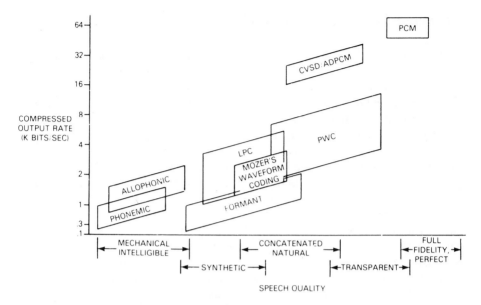

Figure F-4 Speech quality versus bit rate for various coding schemes. (Adapted from Berney and Harshman, 1982.) Reprinted with permission of Cahners Publishing Co.

Speech chips with limited vocabularies are available in the range of $10 and up. To construct the initial representations for new words (to be stored in ROMs) runs upward of tens of dollars per word.

Programming advanced speech synthesizers, to be used with speech generation from text, is an enormous task. The flow diagram for such a state-of-the-art system is given in Fig. F-5. First, the printed text must be converted into phonemes by using a combination of rules and a stored pronouncing dictionary, taking into account pitch, intensity, and duration associated with emphasis, as influenced by word use determined by the syntax of the sentence. The resultant allophones (phonemic variations) are then fed to a phonemic voice synthesizer.

The major commercial application thus far for speech generation from text is reading systems for the blind. These products input text using optical character recognition, and output speech using a text-to-speech synthesizer. Other applications include electronic mail-to-voice, and proofreading.

F-11. COMMERCIAL SYSTEMS

An indication of manufacturers and available commercial systems is given by Table F-2.

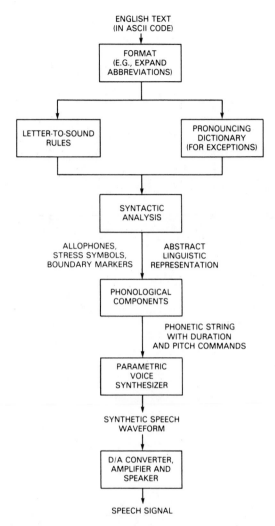

Figure F-5 Text-to-speech conversion. (From Zue, 1982.) Used by permission of Professor V. W. Zue.

F-12. PROBLEMS AND ISSUES

- There is a trade-off in system design between speech quality, vocabulary size, and cost.
- There is the problem as to how best to divide the fundamental units to be used—allophones, syllables, words. The smaller units permit very large vocabularies without excessive storage requirements, while the larger units (such as phrases) provide superior speech quality.

TABLE F-2 Some Commercial Synthesizer Systems (1983)

Manufacturer	Model	Cost	Type	Comments
Votrax (Troy, MI)	VSM/1	$995	Formant	Single-board complete system incorporating a programmable memory
	SVA		Formant	Single-board synthesizer for unlimited text to speech (no internal word storage)
	SC-02		Formant	Synthesizer chip with phoneme library
Speech Plus, Inc. (Mt. View, CA)	Prose 2000	$3500	Formant	Single-board system achieving an unlimited vocabulary capability by using 400 rules and a 3000-word exceptions lexicon; for use with text
Texas Instruments (Dallas, TX)	Speech 1000	$1200	LPC	Synthesizer board with up to 6 minutes stored vocabulary
	TMS 5220	$5	LPC	Single-chip voice synthesizer processor
	TMS 6100	$5	LPC	Single-chip voice synthesizer memory
	Speech synthesizer for TI 99/4A Personal Computer	$100	LPC	Text to speech implemented in 99/4A
	TM 990/306		LPC	Speech module (does not have unlimited vocabulary capability of formant systems)
National Semiconductor (Santa Clara, CA)	Digitalker MM 54104		Mozer's waveform digitizer	Single chip with 256 possible addressable expressions
Centigram (Sunnyvale, CA)	GIM	$350	Formant	SBX module using the G1250 synthesizer chip
	SYBIL	$495	Formant	Single-channel synthesizer for the IBM PC
Kurzweil Computer Products (Cambridge, MA)	Reading Machine for Blind	$30,000	Formant	Uses Speech Plus Prose 2000 synthesizer
American Microsystems (Santa Clara, CA)	53610	a	LPC	
	53620	a	LPC	
General Instruments (Hicksville, NY)	Allophone Synthesis Module		LPC	Annunciates 64 allophones

Manufacturer	Model	Price[a]	Type	Comments
	SP250	a	Formant	Single-channel synthesizer
	SP256	a	Formant	Single-channel synthesizer with microprocessor control
Hitachi	HD 38880	a	PARCOR	Uses partial autocorrelation (closely related to LPC)
Nippon Electric Corp			PARCOR	
Sanyo	LC 1800	a	PARCOR	
Mitsubishi	M58817	a	PARCOR	
Matsushita (Japan)		a	LPC	
Master Specialties (Costa Mesa, CA)	1650	$500 + Vocabulary at $50/word	Word synthesis	
Intex Micro Systems (Troy, NY)	Intex-Talker		Text-to-speech synthesizer	Uses a text-to-phoneme algorithm and a Votrax SC-01 chip
Motorola			CVSD	Encoder and decoder chips
Phillips/Signetics	MEA 8000	a	Formant	
	MEA 10000	a	Formant	
OKI Semiconductor			ADPCM	Encoder and decoder chips

[a] Chip prices range from $3 to $15 depending on model and quantity. Speech Plus provides custom vocabulary generation services for speech synthesizer chips at $100 per word.

- Memory cost considerations tend to restrict the use of the word synthesis approach.
- As the synthesizer techniques improve, it may be that errors due to low sampling rates and inadequate consideration of coarticulation and prosodic (speech stress) effects may be the limiting factors.
- Speech compression techniques are crucial to minimize memory requirements in the synthesizer.
- The high cost of generating words for synthesizer vocabularies needs to be reduced.
- Similarly, the high cost of storing words in ROM needs to be addressed.
- Updating stored vocabularies is problematical due to the need to keep the same speaker available.

F-13. FORECAST

Although the market for voice synthesizers has been relatively small, it is estimated that it will be close to half a billion dollars in the mid-1980s and will reach several billion dollars by 1990. Talking devices will have a big impact on industrial operations, a major effect on learning devices, and will probably be ubiquitous throughout home and consumer products. These devices will be a boon to the handicapped, in everything from talking typewriters and appliances, and reading machines for the blind, to speech prosthetics. It is also anticipated that these devices will be found virtually everywhere in vehicles and transportation systems.

Because of their integration into single chips, the cost of stored vocabulary devices will continue to drop, so that basic hardware costs of less than $10 for units having vocabularies of several hundred words are foreseen by the end of this decade.

REFERENCES

Andreiev, N., "Speech Synthesis: High Technology's Dark Horse in Search of New Pastures," *Control Engineering*, Sept. 1981, pp. 95–98.

Berney, C. L., and Harshman, C., "Voice Ware Does It Differently, *Mini-Micro Systems*, Mar. 1982, pp. 183–193.

Ciarca, S., "Use ADPCM for Highly Intelligible Speech Synthesis," *Byte*, June 1983, pp. 35–49.

Elphick, M., "Talking Machines Aim for Versatility," *High Technology*, Sept.–Oct. 1981, pp. 41–48.

Sherwood, B. A., "The Computer Speaks," *Spectrum*, Aug. 1979, pp. 18–25.

Zue, V. W., *Tutorial on Natural Language Interfaces: Part 2-Speech*. Menlo Park, CA: AAAI, Aug. 17, 1982.

AUTHOR INDEX

SUBJECT INDEX